# Actinomyces, Filamentous Bacteria

## Biology and Pathogenicity

John M. Slack

Mary Ann Gerencser

West Virginia University

Morgantown, West Virginia

**Burgess Publishing Company**
Minneapolis, Minnesota

Copyright © 1975 by Burgess Publishing Company
Printed in the United States of America
Library of Congress Card Number 74-30749
ISBN 0-8087-1993-9

0   9   8   7   6   5   4   3   2   1

TO

Arden Howell, Jr.

A lifelong contributor to the advancement of
knowledge of the actinomycetes

# Contents

# Preface

The purpose of this monograph is to provide both historical and current information about certain bacteria that form filaments which branch. They may cause disease in humans and/or are present in their oral cavity; they include: *Actinomyces, Arachnia, Bacterionema, Rothia* and *Nocardia.*

These organisms have been the subject of controversy since the 1820s from the standpoints of nomenclature, identity and role in human disease. In fact, these same problems still continue; for example, should *Arachnia* be reclassified? Are *A. naeslundii* and *A. viscosus* separate species? What is the role of *Actinomyces* in periodontal disease? Although this monograph does not specifically answer these questions, it is the hope of the authors that the accumulated information will provide a broad base upon which to add new data so that future changes, particularly in nomenclature, can be adequately supported.

Interest, particularly in *Actinomyces*, began when Dr. A. T. Henrici at the University of Minnesota directed the question of pathogenicity of these organisms for laboratory animals to one of the authors (JMS). Then, in 1961, we both became involved in the establishment and work on an international committee which has continued up to the present time. This group was formally established in 1963 as the "Subgroup on Taxonomy of Microaerophilic Actinomycetes" as a division of the "Subcommittee on the Taxonomy of the Actinomycetales" of the International Committee on Nomenclature of Bacteria, International Association of Microbiological Societies. Dr. David Gottlieb was chairman of the Subcommittee. The membership of the Subgroup included: H. Beerens, France, C. S. Cummins, England, L. K. Georg, U.S.A., M. N. Gilmour, U.S.A., P. Holm, Denmark, A. Howell, U.S.A., F. A. Lentze, Germany, E. Meyer, U.S.A., P. Negroni, Argentina, L. Pine, U.S.A., J. M. Porteus, Scotland, G. W. Robertstad, U.S.A., and J. M. Slack, U.S.A., Chairman.

The members of this Subgroup submitted reports of their experimental findings to the chairman, and certain of these data have been included in the various tables compiled for this monograph without benefit of recognition of the particular investigator. However, this was willingly done so that the information could be made available to all persons working with this most interesting but frustrating group of organisms.

# Acknowledgments

This monograph could not have been written without the help of many people, and we wish to express particular thanks to Lucille Georg, John J. Duda, Samuel J. Deal and Pablo Negroni. A number of former students contributed immeasurably to much of the data used throughout the monograph, and these included Sandra Landfried, Jennifer Gossling, Patricia Collins and Jean Ann Setterstrom. The continued secretarial help by Mrs. Lillian Scanga and Mrs. Evelyn Duncil is gratefully acknowledged.

We also wish to credit Phillip Allender, Division of Infectious Diseases, West Virginia Medical Center, for the scanning electron micrograph of the *Arachnia propionica* microcolony (45°, 20KV, X3800) used on the cover.

Much of this work over the years was supported in part by the Public Health Service grants AI-01801 from the National Institute of Allergy and Infectious Diseases and the Institute of Dental Research Grant DE-02675. The grant GB-4909 from the National Science Foundation did much toward making the work of the Subgroup possible.

The enduring patience and support of our respective wife, Jerry, and husband, Vince, are more than appreciated.

# 1

# Introduction

This monograph is designed and written to provide descriptive and pictorial information on the genera *Actinomyces, Arachnia, Rothia, Bacterionema* and *Nocardia.* These bacteria belong to the *Actinomycetales,* and all except *Nocardia* are in the *Actinomycetaceae.* This family includes organisms which form branching filaments at some stage of their growth, although branching may be difficult to demonstrate and many of the organisms more characteristically appear as diphtheroidal forms with occasional long filaments. The production of these branching filaments has in the past been responsible for these organisms being considered fungi, but, in fact, they are procaryotes with a typical bacterial cell wall and are sensitive to antibacterial antibiotics.

Factors which have complicated studies with the actinomycetes include nomenclature, isolation of pure cultures and the maintenance of these cultures. Until recently, the names *A. bovis* and *A. israelii* have been used interchangeably. Now, however, the species can be separately identified, and such confusion should not be perpetuated. The other, and even more important, complication is the isolation of pure cultures, which requires careful picking of single colonies and streaking and repicking of the colony a minimum of three times. After a pure culture is obtained, strains must be checked at frequent intervals to insure continued maintenance of a pure culture. It seems that the actinomycetes have a propensity for becoming contaminated with diphtheroids or staphylococci, and these may be difficult to recognize. The use of such mixed cultures has accounted for reports of excessive strain variability, as well as reports of catalase + or gelatin liquefying *A. bovis* or *A. israelii* (225, 243, 431).

In the following sections of the monograph material is presented on the history, taxonomy and pathogenicity of these various genera and species. Emphasis is placed on morphological, biochemical and serological characteristics of the organisms with this information being presented as much as possible in tabular form. Most of these tables are arranged to provide information on strain variation. It is the hope of the authors that future investigators will add to this data in order to further clarify the taxonomic status of these species or use this data as a baseline for any proposed new genera or species.

In addition, a section on pathogenicity is included. Neither actinomycosis nor nocardiosis is common, although they are not rare, because such infections are reported each year in this country. Also, it is becoming established that various species of *Actinomyces* can play a role in the etiology of caries and periodontal disease, which means they are involved in the most common infectious diseases of man.

# 2

# History

The name *Actinomyces* has been embroiled in controversy since the 1820s, and only relatively recently has there been a semblance of universal agreement on this generic name and in the separation of the various species. This all began when Meyen (316) in 1827 applied the name *Actinomyce horkelli* to a fungus and started the question as to the author of the generic name. It so happened that this fungus had been previously named *Tremella meteorica,* and thus *Actinomyce* was not valid, and Meyen was afforded no claim to the establishment of the name.

In 1854 Graefe (172) described tightly matted fungus masses from concretions taken from a case of canaliculitis of the eye. Then in 1875 Ferdinand Cohn (75) received some concretions from another case of canaliculitis and examining these microscopically described them as containing ". . . smooth, hairlike filaments ramifying and often gently or tightly interlocked, but which show branching if but only sparsely." He called the organism *Streptothrix foersterii,* but this generic name had been preempted by Corda (84) in 1839 and thus was not valid.

In 1876 Bollinger presented a paper discussing tumors of the jaw of cattle, some of which contained granules. He forwarded some of this material to his botanist friend Harz who suggested the name *Actinomyces bovis* for the invading organism. This was published (37) in 1877 and on page 485 it is stated, "Was die Classification dieses neuen Mikroparasiten anlangt, der wegen seines strahlenförmigen Baues nach dem Vorschlage von Harz als '*Actinomyces bovis*' (Strahlenpilz) zu beizeichnen ist . . . ." In 1879 Harz (191), in a rambling article, described at much greater length the granules (with drawings) and filaments as well as his negative attempts at cultivation using a number of substrates, including cooked spring water, bread, plum distillate and cherry extract, with incubation in a warming cupboard. In 1878 Rivolta (387) interpreted the clubs within the granule as disc-shaped lamella and proposed the name *Discomyces bovis* which prompted some arguments, but eventually he withdrew this name in favor of *Actinomyces,* which had priority. See Table 2-1 for a list of names suggested for *Actinomyces.*

Following the unsuccessful attempts of Harz to culture *Actinomyces,* a number of authors either reported negative results or mistakenly labeled their aerobic cultures of *Streptomyces* as *A. bovis.* But in 1889 Bujwid (63) did comparative aerobic and anaerobic cultures (using Buchner's method with 10% alkaline pyrogallic acid) on pus from a fistula of a patient and obtained growth only by the anaerobic method. He shows two photographs of gram-positive branching filaments. These represent the first successful cultivation from man, and it would be assumed that this organism was *A. israelii.* Following this, Wolff and Israel (479) in 1891 published one of the more complete early papers and confirmed the successful cultivation of *Actinomyces* on agar medium under anaerobic conditions. They cultured the granules directly from a cervico-

**Table 2-1. Generic names which have been proposed but not accepted for the name *Actinomyces*.***

| Name | Author, Date |
|------|--------------|
| *Actinomyce* | Meyen, 1827 |
| *Streptothrix* | Cohn, 1875 |
| *Discomyces* | Rivolta, 1878 |
| *Sarcomyces* | Rivolta, 1879 |
| *Actinocladothrix* | Afanassieff and Schulz, 1889 |
| *Nocardia* | Trevisan, 1889 |
| *Oospora* | Sauvageau and Radais, 1892 |
| *Micromyces* | Gruber, 1893 |
| *Cladothrix* | Mace, 1901 |
| *Actinobacterium* | Haass, 1906 |
| *Sphaerotilus* | Engler, 1907 |
| *Cohnistreptothrix* | Pinoy, 1913 |
| *Aёrothrix* | Wollenweber, 1921 |
| *Pionnothrix* | Wollenweber, 1921 |
| *Brevistreptothrix* | Lignierés, 1924 |
| *Corynebacterium* | Haupt and Zeki, 1933 |
| *Proactinomyces* | Negroni, 1934 |

*Most were used with the species names *bovis* or *israelii*. Informative literature includes: Breed and Conn (47), Waksman (468), Lessel (283) and Index Bergeyana (61).

facial and a pulmonary case of actinomycosis and obtained visible growth in 4-5 days. Aerobic cultures from the same material were negative. They also obtained growth in raw or soft-boiled eggs which were pierced with a hot sewing needle and then inoculated with a wire. In addition, rabbits and guinea pigs were inoculated by various routes with the development of varying degrees of abscess formation in which the organisms with clubs were demonstrable. The article is well illustrated with photographs and drawings showing typical actinomycete morphology.

Bostroem (41) also in 1891 wrote an extensive article on his studies of eleven bovine and twelve human cases of actinomycosis. He did aerobic cultures using gelatin or agar plates and isolated filamentous organisms that turned from white to rust-brown or brick-red and which he called "Actinomyces bovis." However, from this description and from his exquisite drawings on Plate IX, these isolates were obviously *Streptomyces*. Bostroem also observed fodder material associated with the bovine infections along with an awn of grain in the center of a lesion and from these postulated that grasses and grains were the "exogenous" source of the actinomycetes causing actinomycosis. Subsequent attempts by other authors (186, 334) to culture *Actinomyces* from vegetation have been uniformly unsuccessful, and it is now assumed that the oral cavity is the "endogenous" source of these organisms. This report by Bostroem introduced the misconceptions that aerobic actinomycetes caused actinomycosis and that vegetable matter was the source of the organism. It has taken some eighty years to correct this misinformation.

In 1919 Breed and Conn (47) recommended that *Actinomyces* be accepted as a genus conservandum with *A. bovis* as the type species. This was accepted by the Judicial Commission and published as such by Winslow et al. (478) in 1920. This opinion was supported by Lessel (283) in his extensive review, and the generic name *Actinomyces* has been consistently used in the recent editions of Bergey's Manual including the eighth edition (419), leading to a worldwide acceptance of the name *Actinomyces* with two exceptions. The French frequently use the name *Actinobacterium* as introduced by Haass in 1906 (179), and the Russians often use *Actinomyces* as the generic name for the aerobic *Streptomyces*. These two misuses continue to add confusion to the literature.

In the preceding section, the establishment of the generic name *Actinomyces* was documented; now there will follow a relatively brief review of the establishment of the species names.

As indicated early in the preceding description, the name *A. bovis* was designated by Harz in 1877 (37), but this species was not actually cultured until 1890 by Mosselman and Lienaux (326). The principal human pathogen (*A. israelii*) was isolated by Bujwid in 1889 (63) and described in considerable detail by Wolff and Israel in 1891 (479). This work was extended by Wright in 1906 (480). There followed many years of confusion involving difficulties in identifying either of the two organisms and the interchangeable use of the names *A. bovis* and *A. israelii*. The differentiation of the two species was shown by the extensive studies of Erikson (115) and the reports of Thompson (455) and Pine et al. (370).

*A. naeslundii* was differentiated from *A. israelii* and described as a separate species by Thompson and Lovestedt in 1951 (456). It was named after Carl Naeslund, who first described such actinomycete-like organisms in the human mouth (332). The closely related *A. viscosus* was first described by Howell in 1963 from periodontal plaque in hamsters, and shortly thereafter he proposed the name *Odontomyces viscosus* (209, 211, 212). This generic name was formulated to indicate habitat and morphology but to separate it from the *Actinomyces* because it was catalase positive (212). Eventually, the description of the genus was changed to include both catalase + and catalase − filamentous organisms, so *Odontomyces* was reclassified as *Actinomyces viscosus* (142).

*A. odontolyticus* was described by Batty in 1958 (17). She isolated over 200 strains from advanced dentinal caries and performed a number of biological tests. It differed from the other species in its greater tolerance for oxygen and the production of a pigmented colony on blood agar. Some additional information on the history of these various species will be found in the succeeding sections of this monograph.

# 3

# Taxonomy
# and Nomenclature

Even though the full name *Actinomyces bovis* was established in 1877, it was not until 1920 (478) that it was conserved by the Judicial Commission. At about this time, there was considerable discussion concerning the classification of bacteria and the establishment of orders and families. Primarily through the efforts of Dr. R. E. Buchanan, the general terminology used in the bacterial classification was established including the Order *Actinomycetales* and Family *Actinomycetaceae*. This Family contained the genera *Actinobacillus, Leptotrichia, Actinomyces* and *Nocardia* (60). There followed then numerous proposals to separate the aerobic from the anaerobic, the gram positive from the gram negative, the branching from the nonbranching, but this was not satisfactorily done until the proposal of Waksman and Henrici (469) in 1943. They maintained the Family *Actinomycetaceae* with only the two genera, *Actinomyces* and *Nocardia*, and established the Family *Streptomycetaceae* for the aerobic filamentous bacteria. The Waksman-Henrici scheme was included in the sixth and seventh editions of Bergey but has been considerably modified in the eighth edition (Table 3-1).

Because of the continuing questions about genera and species to be included in the *Actinomycetaceae* and the need to further establish type and neotype strains of the genera, an international committee entitled "Subgroup on the Taxonomy of the Microaerophilic Actinomycetes" was formalized in 1963 (see Preface for further details and membership of the Subgroup). First, the Subgroup spent considerable time reviewing the literature and establishing uniform procedures prior to having cultures circulated by Dr. E. F. Lessel of the American Type Culture Collection (417). Eventually, particular cultures were selected and deposited in ATCC as suggested neotype strains for the various genera, or if a type strain was already available then its purity and characteristics were verified (Table 3-2). This Subgroup also became actively involved in rewriting and preparing the revised classification of the *Actinomycetaceae* for the eighth edition of Bergey's Manual as listed in Table 3-1. *Bifidobacterium* was studied by another Subgroup, and the descriptions were prepared by M. Rogosa.

The first major change in the revision was the removal of *Nocardia* from the family on the basis that it was aerobic, does not ferment sugars, and contains Dap, arabinose and nocardomycolic acid in the cell wall. It is being placed in the Family *Nocardiaceae* along with the genus *Pseudonocardia*. Then the genus description of *Actinomyces* was modified to include catalase + organisms permitting the inclusion of *A. viscosus* (142), which had originally been described as *Odontomyces viscosus* (212). At the same time, *A. propionicus* (56) was removed from the genus and reclassified as *Arachnia propionica* (365) as it produced propionic acid and had Dap in the cell wall (see description of *Arachnia*).

**Table 3-1.** Family *Actinomycetaceae* adapted from *Bergey's Manual of Determinative Bacteriology,* 8th edition.

Family I. *Actinomycetaceae,* Buchanan 1918

    Genus I. *Actinomyces,* Buchanan 1918
        Species 1. *A. bovis,* Harz 1877, type species of the genus
        Species 2. *A. odontolyticus,* Batty 1958
        Species 3. *A. israelii* (Kruse) Lachner-Sandoval 1898
        Species 4. *A. naeslundii,* Thompson and Lovestedt 1951
        Species 5. *A. viscosus* (Howell, Jordan, Georg and Pine)
                     Georg, Pine and Gerencser, 1969

        Species incertae sedis
           1. *A. eriksonii,* Georg, Robertstad, Brinkman and Hicklin 1965
           2. *A. humiferus,* Gledhill and Casida 1969
           3. *A. suis,* Grasser 1957

        Nomina dubia
           *A. baudetii,* Brion 1942
           *A. discofoliatus,* Grüter 1932
           *A. silberschmidti* (Chalmers and Christopherson) Dodge 1935
           *A. suis,* Gasperini 1892

    Genus II. *Arachnia,* Pine and Georg 1969
        Species 1. *A. propionica* (Buchanan and Pine) Pine and Georg 1969

    Genus III. *Bifidobacterium,* Orla-Jensen 1924
        Species  1. *B. bifidum* (Tissier) Orla-Jensen 1924
        Species  2. *B. adolescentis,* Reuter 1963
        Species  3. *B. infantis,* Reuter, 1963
        Species  4. *B. breve,* Reuter 1963
        Species  5. *B. longum,* Reuter 1963
        Species  6. *B. globosum,* Scardovi et al. 1969
        Species  7. *B. ruminale,* Scardovi et al. 1969
        Species  8. *B. suis,* Matteuzzi et al. 1971
        Species  9. *B. asteroides,* Scardovi and Trovatelli 1969
        Species 10. *B. indicum,* Scardovi and Trovatelli 1969
        Species 11. *B. coryneforme,* Scardovi and Trovatelli 1969

    Genus IV. *Bacterionema,* Gilmour, Howell and Bibby 1961
        Species  1. *B. matruchotii* (Mendel) Gilmour, Howell and Bibby 1961

    Genus V. *Rothia,* Georg and Brown, 1967
        Species  1. *R. dentocariosa,* George and Brown, 1967

    There are three additional species with uncertain status which are described as species *incertae sedis.* The first, *A. eriksonii,* will no doubt be renamed as *Bifidobacterium eriksonii* because its fermentation end products and cell-wall structure are those of the bifidobacteria. The second is *A. humiferus* (160), which morphologically resembles the actinomycetes but differs in that it is an inhabitant of the soil, grows best at 30 C, is sensitive to lysozyme and has a high G + C content (73%). This organism requires additional study and probably the creation of a new genus.

**Table 3-2. Reference strains of the *Actinomycetaceae*.**

| Organisms | Strain Numbers[a] | | Status |
|---|---|---|---|
| **A. Type and Neotype Strains** | | | |
| *A. bovis* | ATCC 13683 | [P1S; WVU 116; CDC X521] | Suggested neotype |
| *A. odontolyticus* | ATCC 17929 | [NCTC 9935; WVU 867; CDC X363] | Cotype |
| *A. israelii* | ATCC 12102 | [H 277; WVU 46; CDC X523; W 855] | Suggested neotype |
| *A. naeslundii* | ATCC 12104 | [H 279; NCTC 10301; WVU 45; CDC X454] | Cotype |
| *A. viscosus* | ATCC 15987 | [H T-6; WVU 745; CDC X603; CDC A828] | Type |
| *A. humiferus* | ATCC 25174 | | Type |
| *A. propionica* | ATCC 14157 | [H 699; ATCC 13682; WVU 471; CDC X364] | Type |
| *R. dentocariosa* | ATCC 17931 | [CDC X599; R X01A; WVU 1089] | Neotype |
| *B. matruchotii* | ATCC 14266 | [NCTC 10254; G 47; WVU 462] | Neotype |
| **B. Additional Reference Strains[b]** | | | |
| *A. bovis* | WVU 292 | | Serotype 2 |
| *A. odontolyticus* | WVU 482 | | Serotype 2 |
| *A. israelii* | WVU 307 | [Holm 8/46-41; CDC W1011] | Serotype 2 |
| *A. naeslundii* | WVU 1522 | [CDC W1516] | Serotype 2 |
| | WVU 1523 | [CDC W1544] | Serotype 2 |
| *A. viscosus* | ATCC 19246 | [WVU 371; N 112; CDC W859] | Serotype 2 |
| *A. propionica* | WVU 346 | [CDC W904] | Serotype 2 |
| *R. dentocariosa* | WVU 477 | [CDC W808] | Serotype 2 |

[a] ATCC = American Type Culture Collection; CDC = Communicable Disease Center; G = Gilmour; H = Howell; NCTC = National Collection of Type Cultures; N = Negroni; P = Pine; R = Roth; WVU = West Virginia University.

[b] Strains used in the WVU laboratory for antiserum production.

The third is *A. suis*, which is complicated. The species name was first used by Gasperini as a synonym for *A. bovis*. Grässer (173) used the name for organisms which he isolated from swine, and, although the organisms were inadequately described and there are no cultures available for study, the name was validly published. Then Biever (28) described swine isolates which he indicated were similar to those of Grässer. More recently Franke (128) described nine swine isolates with adequate morphological and biochemical detail using the name *A. suis* as a new species designation and indicated his strains were somewhat different from strains previously described. Some of these isolates have been examined in this laboratory and by C. S. Cummins with the general agreement that they fit within the characteristics of the *Actinomyces*, but, on the basis of cell-wall and serological studies, they are different from the presently described species. This then presents the problem that: 1) if these three groups of organisms have common identifiable characteristics, then they can be named *A. suis*, Grässer; 2) if *A. suis*, Franke, is considered to be different, then *A. suis*, Grässer, is declared *nomina dubia* because no cultures are available for comparison and a new species name has to be proposed. This question can only be resolved by someone collecting such isolates from sources throughout the world and comparing the biochemical, cell-wall and serological characteristics with published results and either confirming the established name or proposing a new species name.

Numerous species names for the *Actinomyces* have been used in past years, but, with the exception of four, they have not been perpetuated. These four have been declared *nomina dubia* because cultures do not exist and the descriptions were inadequate to permit identification; they are: *A. baudetii*, *A. discofoliatus*, *A. silberschmidti* and *A. suis*, Gasperini.

In devising the presently proposed classification of the *Actinomyces*, an essentially phenetic approach was used. For the most part the available data did not lend itself to a numerical taxonomic approach, but an attempt was made to determine the greatest overall similarity rather than using single tests as discriminative criteria. Holmberg and Hallender (207) did a numerical study of species of *Actinomyces*, *Rothia*, *Bacterionema* and related bacteria indicating that the genera did individually cluster. *A. naeslundii* and *A. viscosus* composed a single cluster, and the few strains of *A. israelii* and *A. propionica* composed a second cluster. This study has the shortcomings that only a few strains were used and that the battery of tests included a large proportion of morphological tests which tend to weight similarities. Unfortunately, genetic information on the *Actinomyces* is quite scant, and the mechanisms of gene transfer are unknown. The available data on DNA base ratios are recorded in Table 3-3. It is interesting to note that the present information supports the combination of *A. naeslundii* and *A. viscosus* and in one case (86) suggests that the hamster isolates of *A. viscosus* (59-60.9% GC) may be different from the human isolates (64.5-69.9% GC) since a difference of 5% in GC ratios implies a species difference (226). However, a second study (183) found the same % GC in one animal and three human isolates. There is no information on DNA homologies on any of the *Actinomyces*, and this is a fertile area for investigation.

Table 3-3. DNA base composition (moles % GC) of *Actinomycetaceae*.

| Organism | Strain Designation | % GC[a] | Reference |
|---|---|---|---|
| *A. bovis* | C35 | 63 | 225 |
| | C274 (ATCC 13683)[b] | 63 | 225 |
| | ATCC (19009) | 57 | Hammond[c] |
| | ATCC 13683[b] | 53.5 | Hammond[c] |
| | ATCC 19013 | 63 | Hammond[c] |
| *A. odontolyticus* | ATCC 17929[b] | 62 | 419 |
| | ATCC 17929[b] | 62 | Hammond[c] |
| *A. israelii* | ATCC 12102[b] | 60 | 419 |
| | ATCC 12836 | 57 | Hammond[c] |

| | | | |
|---|---|---|---|
| | B5 (ATCC 12103) | 65 | Hammond[c] |
| | B6 (ATCC 12597) | 65 | Hammond[c] |
| | A7 (ATCC 12102)[b] | 63 | Hammond[c] |
| | A8 (ATCC 10048) | 63 | Hammond[c] |
| *A. naeslundii* | W1544 | 68.5 | 86 |
| | I | 64 | Hammond[c] |
| | ATCC 12104[b] | 63 | Hammond[c] |
| *A. viscosus* | ATCC 15987[b] | 59 | 419 |
| | ATCC 15987[b] | 60.9 | 86 |
| | M100 | 64.5 | 86 |
| | 5-5S | 66.3 | 86 |
| | ATCC 19246 | 67.4 | 86 |
| | M112 | 68.4 | 86 |
| | Scheiner 1 | 68.6 | 86 |
| | R28 | 69.9 | 86 |
| | T6 (ATCC 15987)[b] | 63 | Hammond[c] |
| | ATCC 19246 | 63 | Hammond[c] |
| | R C45 | 64 | Hammond[c] |
| | T14 | 63 | Hammond[c] |
| *A. humiferus* | 12 strains (not numbered) | 73 | 160 |
| *Arachnia propionica* | 0026[b] | 64 | 225 |
| | ATCC 14157[b] | 65 | 225 |
| | 5068 | 64 | 225 |
| | 5072 | 65 | 225 |
| | 5074 | 63 | 225 |
| | A3 (ATCC 14157)[b] | 60.5 | Hammond[c] |
| *Rothia dentocariosa* | ATCC 17931[b] | 51 | Hammond[c] |
| | ATCC 14190 | 50 | Hammond[c] |
| | ATCC 14189 | 47 | Hammond[c] |
| | ATCC 14191 | 49 | Hammond[c] |
| *(N. salivae)* | ATCC 19426 | 53 | Hammond[c] |
| *Bacterionema matruchotii* | several | 55-58 | 351 |
| | 47 (ATCC 14266)[b] | 52.5 | Hammond[c] |
| | 100 | 50 | Hammond[c] |
| | 102 | 53 | Hammond[c] |
| | 208 | 53 | Hammond[c] |
| | 214 | 56 | Hammond[c] |
| *(L. dentium)* | ATCC 19419 | 55 | Hammond[c] |

[a] Values obtained by thermal denaturation.
[b] Type, neotype or proposed neotype strain.
[c] Unpublished data courtesy of B. F. Hammond.

The classification as presented in the eighth edition of Bergey's Manual is an advancement over previous editions, and it provides adequate information for the identification of most clinical isolates and many of those present as normal oral flora. However, any classification scheme is of necessity subject to change as additional data accumulates. Hopefully, however, such changes will only be made after an adequate amount of information has been obtained from the study of a number of strains (ten and preferably more). It is without question the right of investigators to publish and discuss data, but it is to be recommended that they refer to such isolates as "X" or as a biotype or by a number or by some trivial designation before attempting to formally propose a new generic or species name, because, once a name is published, it is very difficult to change or nullify (the present situation with *A. suis* is a good example). A revision of the present classification should be based on a large number of different phenotypic characteristics, DNA homologies and additional genetic information.

Scharfen (405, 406) recently proposed a redefinition of the genus description to include organisms containing Dap, the basis being that he described a strain labeled Trutnov 139/66 which biochemically resembles *A. israelii* but produces aerial filaments and presumably has two different types of peptidoglycan: combination 1, containing lysine +, ornithine −, LL-Dap +, and combination 2, containing lysine +, ornithine +, LL-Dap −. He quotes the report of Sukapure et al. (445) in which it was indicated *A. israelii* 12102 contained LL-Dap in the cell wall. We obtained this particular culture from M. Lechevalier and found that the culture used was a mixture of *A. israelii* and *P. acnes* with the *P. acnes* accounting for the presence of the Dap.

This does not necessarily explain the report of Scharfen because we have not studied his isolate; however, a number of laboratories have done cell-wall analysis with pure cultures of *A. israelii* and Dap has not been reported. Again, we would like to reemphasize the necessity of purifying cultures through repetitive, single-colony transfers and then frequently checking the purity of the isolate.

We have not observed the aerial filaments as reported by Scharfen in strain 12102, but we have two oral isolates which are typical *A. israelii*, serotype 1, that produce numerous and distinct aerial filaments, but they do not contain Dap (see Chapter 6, Figure 6-49). Thus, this colonial morphological change can occur, but the possibility that this is accompanied by a change in the composition of the peptidoglycan will have to wait further study and confirmation.

The possible relationship of *Corynebacterium pyogenes* to *Actinomyces* should be mentioned. *C. pyogenes* biochemically is similar to the actinomycetes except it is actively proteolytic. The microcolonies and macrocolonies on BHI resemble those of *A. bovis* and *A. odontolyticus*, and a low-titered serological cross-reaction is demonstrable between *C. pyogenes* and *A. odontolyticus* with FA. The G + C base ratio is 57%, which is within the 50-65% of the *Actinomyces* but considerably above that of most streptococci. On the basis of cell-wall studies by Cummins and Harris (91) and serological cross-reactions between extracts of *C. pyogenes* and group G streptococcus antiserum, the suggestion has been made to consider the organism a *Streptococcus*. We are not recommending that *C. pyogenes* be classified as an *Actinomyces*, but this possibility should be considered in the reassessment of the taxonomic position of this organism.

The three additional monotypic genera within the *Actinomycetaceae* included in the monograph are *Arachnia*, *Rothia* and *Bacterionema*. The genus *Arachnia* (spiderlike) was established (365) to accommodate the organism originally described as *Actinomyces propionicus* because its fermentation end products (propionic acid) and Dap in the cell wall were not compatible with the other species of *Actinomyces*. At present the question of its relationship to the other propionic-acid bacteria remains, and it has been suggested that *A. propionica* be moved to the genus *Propionibacterium*. Johnson and Cummins (225) showed high DNA homologies among strains of *A. propionica* but low homologies with various species of *Propionibacterium* including *P. acnes*. On the basis of its actinomycete morphology including typical spiderlike microcolonies and its ability to cause clinically typical actinomycosis, it could remain in the *Actinomycetaceae*. If moved to another family, it should retain the generic name *Arachnia*.

The genus *Bacterionema* was created (157) to separate the organism from *Leptotrichia buccalis* primarily on the basis of its being aerobic and having gram-positive branching filaments with a characteristic "whip-handle" structure (see Chapter 9 on *Bacterionema* for further details). Only one species has been described, but studies have indicated considerable variability in oxygen requirements, and the more anaerobic strains may be an additional species. The question has been raised as to whether or not *Bacterionema* belongs in the *Actinomycetaceae* as it has both m-Dap and arabinose in the cell wall, produces propionic acid as an end product from glucose fermentation and has a somewhat lower G + C base ratio (55-58%). The question is justified, but at present there is no obviously better family, although it shows some affinity for the *Corynebacteriaceae*.

This brings up the point that many of the classification problems center around the placement of an organism in a particular family which really has little to do with identification; thus it might be well to eliminate the family concept and concentrate on well-defined and recognized genera which, for convenience, could be placed in broadly defined groups.

The genus *Rothia* was established for the organism previously named *Nocardia dentocariosa* or *Nocardia salivae*. *Nocardia* was an inappropriate generic name because *Rothia* ferments glucose and contains lysine rather than Dap in the cell wall (139). Morphologically, it resembles the actinomycetes but has a peptidoglycan structure not found in other actinomycetes (409). The majority of the strains described in the literature were surprisingly uniform in their characteristics, thus supporting the description of the single species *R. dentocariosa*. However, the study of a larger number of isolates from the oral cavity in this laboratory indicates biochemical and serological differences and suggests that some strains may represent a new species. This requires further study of more isolates and the collection of genetic information. This genus is firmly established and fits well into the *Actinomycetaceae* (see Chapter 8 on *Rothia* for further details).

# 4

# Physiology of the
# *Actinomycetaceae**

Within the genus *Actinomyces,* the species and even strains within a species vary in their ability to grow anaerobically or aerobically. The cause for this variation is unknown, but the explanation must be related to the metabolic processes of the organism. The main metabolic investigations have been performed by Buchanan and Pine (58) working with strain 10049 of *A. israelii,* an organism called *A. naeslundii* in the publications by these investigators. Under the conditions used in the Buchanan-Pine study, this bacterium produces energy by substrate phosphorylation with the amount of ATP produced dependent on cultural conditions. Anaerobically without added $CO_2$, the fermentation is homolactic. In the presence of $CO_2$, the fermentation is heterolactic with formate, acetate, lactate and succinate as products. Growth is more rapid and cell yield is greater with $CO_2$.

In air, small amounts of $CO_2$ are required, and acetate and carbon dioxide are the main products of glucose catabolism. Growth studies indicate that four ATP's are formed per mole of glucose (59) probably from the phosphoroclastic split of pyruvate with production of acetyl phosphate. Only two ATP's were formed when growth was anaerobic without $CO_2$, i.e., a homolactic fermentation. They also demonstrated fructose—1,6-diphosphate aldolase and phosphate acetyltransferase in cell-free extracts.

One of the most interesting recent discoveries in *Actinomyces* is the presence of heme proteins resembling cytochrome b (451). *A. bovis, A. viscosus* and eight strains of serotype 1 of *A. israelii* showed absorption maxima of 559 when grown aerobically. Serotype 2 of *A. israelii* did not show such absorption, and growth under anaerobic conditions diminished the absorption of all the organisms except *A. bovis. A. viscosus* gave evidence of other cytochrome peaks, possibly a and c. The function of the cytochromes has not been studied, but a role in energy production is possible. It was also noted that the detection of the cytochromes by low-temperature spectrophotometry did not correspond with results obtained using the benzidine test on colonies. *Actinomyces viscosus,* although related to other species of *Actinomyces,* is aerobic and contains catalase.

Nutrition of growth has been the object of a few investigations with *Actinomyces* (70, 214, 238). Except for a few strains, synthetic media have not been developed for these bacteria. The complexity of these media indicates, however, that in general the *Actinomyces* possess limited anabolic capacity since most of the amino acids and B vitamins must be added. In addition, certain growth factors such as purines and pyrimidines are inhibitory to growth (238). Perhaps

*Prepared by S. J. Deal, WVU.

imbalance in nutrients rather than a deficiency in nutrients is the main problem in developing synthetic media for these bacteria.

*Arachnia propionica* ferments glucose to carbon dioxide, acetate and propionate with small amounts of succinate and lactate.

The genus *Rothia* is an aerobe which produces lactic acid when growing on glucose (211). Carbon recoveries were low, and it is assumed that carbon dioxide, which was not measured, was produced.

*Bacterionema* grows anaerobically on glucose with concomitant reduction in pH (155, 215). Formic, acetic, lactic, propionic and succinic acids were detected in varying concentrations with lactate formed in greatest concentration. Carbon dioxide was also produced. Better growth was obtained aerobically, and only a few volatile acids (but no lactate) were produced. Unlike *Actinomyces,* growth of *Bacterionema* was not affected by carbon dioxide. A synthetic growth medium for two strains of *B. matruchotii* has been prepared by Takazoe (448) which contains eleven to thirteen amino acids, eleven vitamins and cofactors, four purines and pyrimidines, pimelic acid and inorganic salts. Glucose was the energy source. *B. matruchotii* contains dehydrogenases for malate, fumarate, succinate and NADH (432). Cytochrome oxidase was not demonstrated, but oxidation of succinate or NADH reduced added cytochrome c. An NADP-dependent iso-citrate dehydrogenase was also demonstrated. These data suggest the presence of a TCA cycle, but further enzyme and respiration studies are necessary to determine the complete cycle.

It is impossible to define specifically the total metabolic capabilities of *Actinomyces* and related genera until many more strains have been examined by many techniques and procedures. The role of carbon dioxide for growth has not been critically explained, and nothing is known regarding biosynthesis. Existing somewhere between anaerobic life and aerobic life, these bacteria should offer an interesting subject for the investigator.

# 5

# Cell Walls of the
# *Actinomycetaceae*

## PEPTIDOGLYCAN

Beginning with the studies by Cummins and Harris (91, 92), there have been numerous investigations into the chemical composition of the cell walls of *Actinomyces,* especially *A. israelii* and *A. bovis:* Cummins (89, 90), Subgroup report (9), Snyder et al. (436), Pine and Boone (363), Boone and Pine (40), DeWeese et al. (97), Lechevalier et al. (272), Georg et al. (143), Hammond (182), Hammond et al. (185), Reed (382), Reed and Evans (383), MacLennan (296), Schleifer and Kandler (409). These studies, summarized in Table 5-1, conclude that the genus *Actinomyces* consists of organisms which have a typical bacterial peptidoglycan with lysine but no diaminopimelic acid (Dap) or arabinose. However, there are a number of conflicting results in these reports involving both the amino acid and sugar composition of the walls. Some of these discrepancies are due to the use of impure wall preparations. For example, we now know that at least some of the walls used in our quantitative study (97) contained cytoplasmic contamination despite the fact that negatively stained preparations examined in the electron microscope appeared clean. This discussion of the peptidoglycan components of *Actinomyces* cell walls will be based primarily on the review by Schleifer and Kandler (409) which reports both quantitative wall composition and possible peptidoglycan structure.

Schleifer and Kandler classified peptidoglycans into two major groups, A and B, based on the position of the cross-linkage between peptide subunits. The major groups were then subdivided on the basis of the type of cross-link.

Two peptidoglycan types were found in the walls of *Actinomyces.* The first type, belonging to their subgroup $A_4$, is found only in *A. bovis* and contains a typical L-alanine, D-glutamic acid, L-lysine, D-alanine sequence in the peptide moiety and a D-aspartic acid cross-link between L-lysine and D-alanine. This peptidoglycan is also found in many lactobacilli and some bifidobacteria.

*A. israelii, A. naeslundii, A. viscosus* and *A. odontolyticus* all belong to group B. In these species, muramic acid, glucosamine, alanine, glutamic acid, lysine and ornithine are in molar ratios of 1:1:2:2:1:1. This molar ratio has *not* been found in any other bacteria, so far, making the amino acid composition of these four *Actinomyces* species unique. Preliminary results on its structure indicate that the peptidoglycan has an interpeptide bridge consisting of glutamyl-ornithine between the alpha carboxyl of glutamic acid and the C terminus of D-alanine (Figure 5-1). L-lysine does not take part in the cross-linkage. Reed (382) analyzed cell-wall preparations of one strain of *A. viscosus* and reported glucosamine, muramic acid, alanine, glutamic acid and

**Table 5-1.** Cell-wall composition of the *Actinomycetaceae*[a].

| | Major Amino Acids[b] | | | | | | Major Sugars | | | | | | |
|---|---|---|---|---|---|---|---|---|---|---|---|---|---|
| | Lysine | Aspartic Acid | Ornithine | LL Dap | DL Dap | Glycine | Glucose | Galactose | Rhamnose | 6-deoxytalose | Fucose | Mannose | Arabinose |
| *Actinomyces* | | | | | | | | | | | | | |
| A. bovis | +[c] | + | – | – | – | – | – | – | + | + | + | – | – |
| A. israelii | + | – | + | – | – | – | – | + | – | – | – | – | – |
| A. naeslundii | + | – | + | – | – | – | + | – | + | + | – | + | – |
| A. viscosus | + | – | + | – | –` | – | + | + | + | + | – | + | – |
| A. odontolyticus | + | – | + | – | – | – | + | + | – | – | – | + | – |
| A. humiferus | + | + | + | – | – | – | – | – | + | – | – | – | – |
| A. suis | + | – | + | – | – | – | – | – | + | – | – | – | – |
| *Rothia* | | | | | | | | | | | | | |
| R. dentocariosa | + | – | – | – | – | – | – | + | – | – | – | – | – |
| *Arachnia* | | | | | | | | | | | | | |
| A. propionica | | | | | | | | | | | | | |
| cell wall type 1 | – | – | – | + | – | + | – | + | – | – | – | – | – |
| cell wall type 2 | – | – | – | + | – | + | + | + | – | – | – | – | – |
| *Bacterionema* | | | | | | | | | | | | | |
| B. matruchotii | – | – | – | – | + | + | + | + | – | – | – | – | + |

[a] Data from Cummins (89, 90, personal communication); Cummins and Harris (91); Schleifer and Kandler (409); Hammond et al. (185); Davis and Freer (95); Johnson and Cummins (225); Gledhill and Casida (160); Davis and Baird-Parker (94).

[b] All walls contain glucosamine, muramic acid, alanine and glutamic acid.

[c] + = present; – = absent.

lysine with molar ratios of 1:1:4:4:4. The ratio of D:L alanine was 1:3. No ornithine was found. On this basis a cross-bridge structure containing equimolar amounts of glutamic acid and lysine (3:3) with two moles of L-alanine was proposed. Reed and Evans (383), using partial acid hydrolysis, obtained three peptides from the peptidoglycan. The first component contained equimolar amounts of muramic acid, alanine and glutamic acid; the second equimolar amounts of alanine, lysine and glutamic acid; and the third contained lysine, alanine and glutamic acid in a ratio of 2:1:1. Without proposing a specific sequence, the authors again suggest a unique cross-bridge component in *A. viscosus* walls with a highly cross-linked structure.

The quantitative and structural studies of Schleifer and Kandler differ widely from those of Reed (382) and Reed and Evans (383) in several respects: 1) in the actual components found, 2) in the molar ratios of the components and 3) in the proposed peptidoglycan structure. These are not strain differences since the same organism, *A. viscosus,* T6 (ATCC 15987), was used in both cases. It is impossible to make a decision concerning the two proposed structures without

**Figure 5-1. Structure of the cell-wall peptidoglycan of the *Actinomycetaceae*[a].**

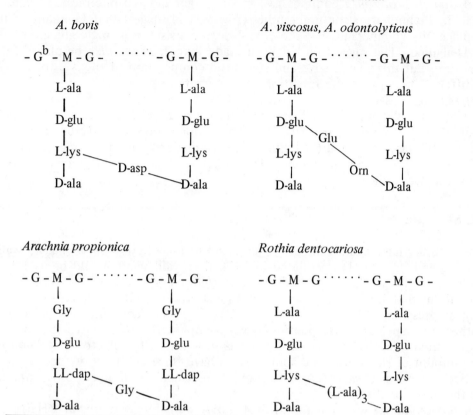

*Actinomyces*

A. israelii, A. naeslundii

A. bovis

A. viscosus, A. odontolyticus

*Arachnia propionica*

*Rothia dentocariosa*

[a]Data taken from Schleifer and Kandler (409).

[b]Abbreviations: G=N–acetylglucosamine; M=N–acetylmuramic acid; ala=alanine; asp=aspartic; dap=diaminopimelic acid; glu=glutamic acid; gly=glycine; lys=lysine; orn=ornithine.

further study. However, it is obvious from these studies that the wall of *Actinomyces* differs from most other gram-positive bacteria, and the four species, other than *A. bovis,* form a closely related group.

Two additional species may be included in the *Actinomyces, A. humiferus* and *A. suis. A. humiferus* (160) has alanine, glutamic acid, lysine, ornithine and aspartic acid in the wall. Two of the strains of *A. suis* described by Franke (128) have been studied by Cummins (personal communication), who found alanine, glutamic acid, lysine and ornithine in the wall. The cell-wall composition of both of these provisional species is compatible with their classification as *Actinomyces.*

Three additional genera to be considered are *Arachnia, Bacterionema* and *Rothia.* Of these, *Rothia* is most like the *Actinomyces* in wall composition. The walls of *Rothia* were studied by Davis and Baird-Parker (94), Georg and Brown (139), Sukapure et al. (445), Hammond (182) and Schleifer and Kandler (409). There are some minor discrepancies in these reports. Davis and Baird-Parker (94) and Georg and Brown (139) reported that the major amino acids are alanine, glutamic acid and lysine, while Sukapure et al. (445) found aspartic acid in addition to these. Schleifer and Kandler (409) studied the same two strains used by Sukapure et al. (445) and found alanine, glutamic acid and lysine but no aspartic acid. Preliminary results showed that the peptidoglycan belongs to the L-lysine-L-alanine$_3$ type A3a (Figure 5-1). This type of wall is common in bacteria but has not been found in any other actinomycetes.

*Arachnia* contains LL-Dap with major amounts of alanine, glutamic acid and glycine in the peptidoglycan (225, 362, 366, 409). Aspartic acid has been reported (40). Schleifer and Kandler (409) showed that *A. propionica* contains two moles of glycine and one mole of alanine per mole of glutamic acid. Preliminary structural studies (Figure 5-1) suggest that L-alanine in the peptide subunit is replaced by glycine and that the cross-bridge consists of a single glycine between LL-Dap and D-alanine. The suggestion has been made that *A. propionica* be placed in the genus *Propionibacterium,* but the quantitative and structural composition of the wall of *Arachnia* is sufficiently different to support the retention of a separate genus.

*Bacterionema* contains Dap in its cell walls, but in this case it is the DL isomer. Davis and Baird-Parker (93) analyzed the cell walls of three strains of *Leptotrichia dentium,* which would now be classified as *Bacterionema matruchotii.* The major amino acids were alanine, glutamic acid and DL- or DD-Dap. Baboolal (10) studied six strains and found alanine, glutamic acid and DL-Dap with moderate amounts of glycine. The presence or absence of glycine needs to be determined by further studies.

## CELL-WALL SUGARS

Only a few studies have dealt with the sugar composition of *Actinomycetaceae* walls even though Cummins and Harris (91) postulated in 1958 that wall sugars would be useful for species differentiation. The major sugars reported in the *Actinomyces* are shown in Table 5-1. The wall of *A. bovis* contains rhamnose, 6-deoxytalose and most strains contain fucose (92). Glucose and mannose may be present in trace amounts (91, 362). *A. israelii* contains only galactose in its wall. The other three species whose peptidoglycan is apparently identical to that of *A. israelii* either do not contain galactose or have other sugars in addition to galactose. *A. humiferus* contains a major amount of rhamnose, and some strains have trace amounts of fucose and glucose (160). The two strains of *A. suis* studied (Cummins, personal communication) contained only rhamnose as the major wall sugar.

The sugar 6-deoxytalose was first found in *A. bovis* walls by Cummins and Harris (92) and identified by MacLennan (296). Hammond et al. (185) found 6-deoxytalose in *A. naeslundii, A. viscosus* and *A. bovis* but not in *A. israelii* or *A. odontolyticus.* They reported that this sugar was a prominent component of the walls of fresh dental isolates resembling *A. naeslundii* and *A. viscosus* and suggested that this was a major antigen. However, Reed (382) found only rhamnose as a major neutral sugar in *A. viscosus* and indicated that it was not an antigenic determinant as the peptidoglycan was the major cell-wall antigen.

Sukapure et al. (445) found galactose and mannose but not arabinose in the walls of two strains of *Rothia.* Hammond (181, 182, 185) reported that galactose was the major sugar of *Rothia* with smaller amounts of glucose, fructose and ribose but no 6-deoxytalose. In a soluble cell-wall antigen, fructose was the major component (60%) and the determinant of serological specificity.

Buchanan and Pine (56) and Boone and Pine (40) reported glucose and galactose in the walls of *Arachnia propionica.* Then, on the basis of cell-wall sugars (Table 5-1), Johnson and Cummins (225) placed *A. propionica* into two groups, one of which resembled *A. israelii* in containing only galactose as a major sugar. The other group contained both glucose and galactose with a trace of mannose.

*Bacterionema* in contrast to the other genera contains arabinose as a major wall sugar. Glucose is also present, but reports vary on the occurrence of galactose, which was found by Baboolal (10) and Pine and Boone (363) but not by Davis and Baird-Parker (94).

Most of the above cell-wall work was done in relation to taxonomy or serological studies. Unfortunately, these studies suffer from the lack of good quantitative data and from the inclusion of only a few strains. However, the results indicate that the species of *Actinomyces,* other than *A. bovis,* are closely related. On the basis of cell-wall composition, *Arachnia* and *Bacterionema* could be removed from the family, but there is no good clue as to where they should be placed.

# 6

# Genus *Actinomyces*

## A. MORPHOLOGY

**Cell Morphology.** All species of *Actinomyces* are gram positive, non-acid-fast, nonmotile and nonsporeforming. They occur as rods and filaments which are less than 1 $\mu$m in width but which vary considerably in length. The short rods (Figures 6-1, 6-2, 6-4, 6-5, 6-8, 6-10, 6-13) are usually in diphtheroidal arrangements including palisades, Y, V and T forms. The filaments, which may be straight or wavy (Figures 6-3, 6-5, 6-6, 6-9, 6-11, 6-12, 6-14), are usually long and slender and may have swollen, clubbed or clavate ends. True branching (Figures 6-3, 6-6, 6-9, 6-11, 6-12, 6-14) occurs in all species, although at times it may be difficult to demonstrate. The relative preponderance of diphtheroidal rods or filaments depends on a number of factors including cultural conditions along with strain and species variation. In gram-stained smears, the staining may be deep and uniform (Figures 6-5, 6-6, 6-12, 6-13) or irregular, sometimes producing a beaded or barred appearance (Figures 6-1, 6-3, 6-7, 6-11). Indications of a granular cytoplasm which may account for the irregular staining can be seen in unstained cells observed in darkfield (Figure 6-8).

In *A. bovis* (Figures 6-1 and 6-2) diphtheroidal rods usually predominate. Short branching rods occur, but long multibranched filaments are seldom observed. Occasionally, rough (filamentous) strains of *A. bovis* are isolated which do show such branching filaments in stained smears. In contrast to the situation in culture, branching filaments are usually seen *in vivo* in both natural and experimental infections with *A. bovis*.

The other four species of *Actinomyces* generally show long branching filaments as well as diphtheroidal forms. *A. odontolyticus* (Figures 6-3 and 6-4) is most like *A. bovis* in that long branching filaments are rare. *A. israelii* and *A. naeslundii* frequently exhibit filaments, and many strains are predominantly filamentous or an obvious mixture of the two forms (Figures 6-9 and 6-11). However, even in these species, diphtheroidal forms may predominate. Howell (209) described the original hamster isolates of *A. viscosus* as long, occasionally sinuous filaments but found that prolonged laboratory cultivation often resulted in completely diphtheroidal cultures. The canine, pig and goat isolates of *A. viscosus* described by Georg et al. (140) consisted of filaments, diphtheroidal forms, rods and cocci. One or more of these forms might predominate in a single culture, but the same form did not necessarily persist in successive subcultures. In our experience, fresh human isolates resemble the animal isolates in morphologic variation. Some human isolates were predominantly diphtheroidal on primary isolation.

Erikson (115) very effectively used an agar slide culture method for observing the typical filamentous morphology of the *Actinomyces*. She noted a predominance of short diphtheroidal

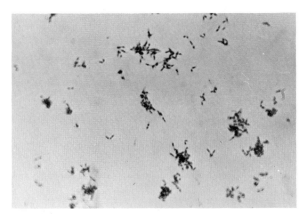

**Figure 6-1.** *A. bovis* gram stain, 48 hr., Thioglycollate Broth, X1200.

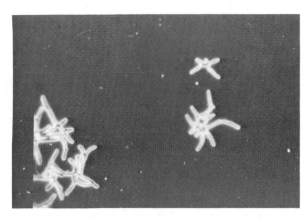

**Figure 6-2.** *A. bovis,* darkfield, 48 hr., Thioglycollate Broth, X1200.

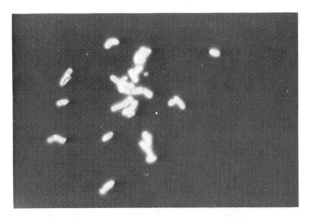

**Figure 6-3.** *A. odontolyticus,* gram stain, 48 hr., Thioglycollate Broth, X1200.

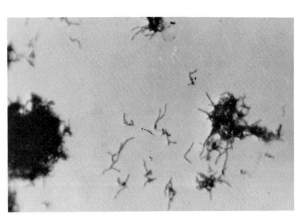

**Figure 6-4.** *A. odontolyticus,* darkfield, 48 hr., Thioglycollate Broth, X1200.

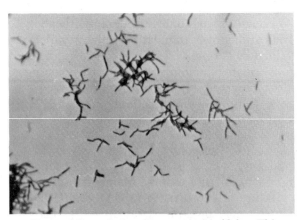

**Figure 6-5.** *A. israelii,* gram stain, 48 hr., Thioglycollate Broth, X1200.

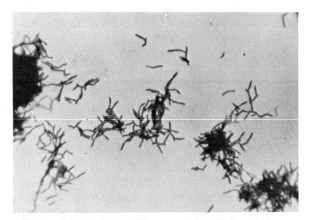

**Figure 6-6.** *A. israelii,* gram stain, 48 hr., Thioglycollate Broth, X1200.

forms in stained smears and indicated that these resulted from the breakup of the delicate filaments during the preparation of the smears. This may be true in some instances, but frequently, even in broth, the diphtheroidal forms predominate.

In carefully prepared smears from broth cultures, most strains of *A. israelii* (even those which emulsify easily) will show groups of filaments resembling a small microcolony (Figures 6-6 and 6-7). Similar microcolonies are seen in cultures of *A. naeslundii* (Figure 6-12) and *A. viscosus.* These groups usually contain intertwining branching filaments, some of which may have bulbous ends.

22  **Genus** *Actinomyces*

Figure 6-7.   *A. israelii,* gram stain, 48 hr., Thioglycollate Broth, X1500.

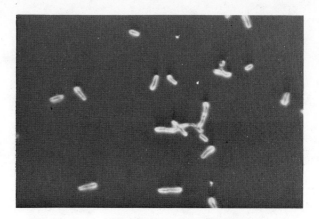

Figure 6-8.   *A. israelii,* darkfield, 48 hr., Thioglycollate Broth, X1200.

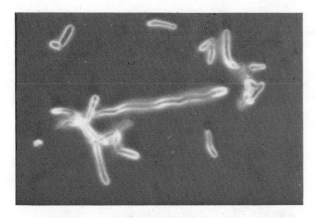

Figure 6-9.   *A. israelii,* darkfield, 48 hr., Thioglycollate Broth, X2400.

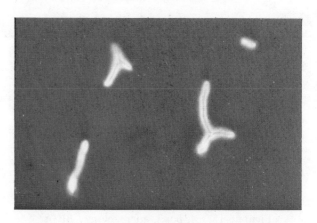

Figure 6-10.   *A. israelii,* darkfield, 48 hr., Thioglycollate Broth, X2400.

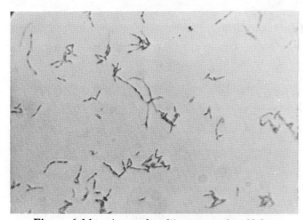

Figure 6-11.   *A. naeslundii,* gram stain, 48 hr.,
Thioglycollate Broth, X1200.

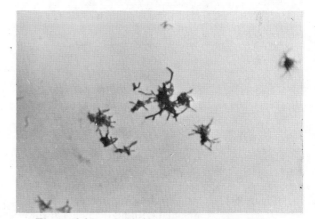

Figure 6-12.   *A. naeslundii,* gram stain, 48 hr.,
Thioglycollate Broth, X1200.

If filaments are not readily seen in a stained smear, a portion of the growth removed from a broth culture with a capillary pipette, crushed gently under a coverslip and examined in darkfield, will often show the branching filaments.

The morphology of *Actinomyces* species has been described by Erikson (115), Howell et al. (213), Coleman et al. (78), Slack and Gerencser (422, 423), Pine et al. (370), Howell (209), Georg et al. (143) and Batty (17).

**Colony Morphology.** The morphology of both micro- and mature colonies has been used in

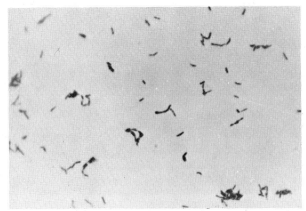

**Figure 6-13.** *A viscosus,* gram stain, 48 hr., Thioglycollate Broth, X1200.

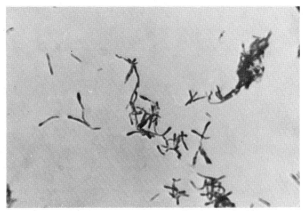

**Figure 6-14.** *A. viscosus,* gram stain, 48 hr., Thioglycollate Broth, X1200.

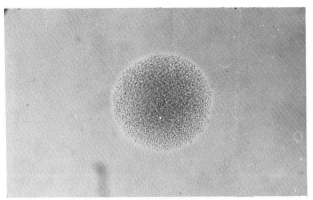

**Figure 6-15.** *A. bovis,* microcolony, BHIA, 24 hr., X330.

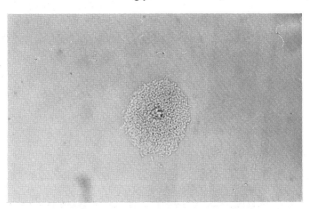

**Figure 6-16.** *A. bovis,* microcolony, with optically dark center, BHIA, 24 hr., X330.

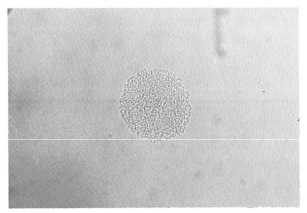

**Figure 6-17.** *A. odontolyticus* microcolony, granular with no filaments at the edge, BHIA, 24 hr., X720.

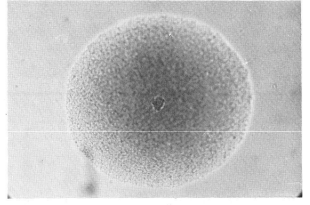

**Figure 6-18.** *A. odontolyticus,* microcolony, granular with optically dark center, BHIA, 24 hr., X330.

*Actinomyces* studies for: 1) differentiation between human and bovine types (115, 455, 480), 2) differentiation between *A. israelii, A. bovis* and *A. naeslundii* (370) and 3) detection of and identification of *Actinomyces* on isolation plates. Each species possesses a typical or most common type of colony at both the microcolony (18-24 hr.) and mature colony (7-14 days) stages. However, it is important to recognize that on any given plate or with a particular isolate there may be a range of colony morphologies so the presence or absence of "typical" colonies cannot be depended upon exclusively for either isolation or identification. In the descriptions which follow, we will try to describe both the typical colonies and the expected range of variation of colony type for each species.

**Microcolony Morphology.** Microcolonies are observed on Brain Heart Infusion Agar (BHIA) plates incubated anaerobically for 18-24 hours (see Chapters 13 and 14 for details). Variations in medium and incubation conditions may affect the microcolony morphology. Under these conditions, *A. bovis* colonies have a smooth or finely granular surface, entire to irregular edge, and are slightly raised to convex, white and soft (Figure 6-15). The colonies may have an optically dark center spot (Figure 6-16). Typical *A. bovis* microcolonies may have irregular, jagged edges but do not show true filaments; however, Thompson (455) described young microcolonies with short radiating filaments at the edge which took on the appearance of typical *A. bovis* microcolonies as they increased in size. Truly filamentous strains of *A. bovis,* usually referred to as rough strains, have been isolated (143, 370). These strains produce small microcolonies with radiating filaments at the edge like the microcolonies of *A. israelii.* Two rough strains lacked cell-wall aspartic acid which is found in smooth-colony strains (363). This is the only reported instance in the *Actinomyces* where a change in morphology correlates with a change in cell-wall composition.

Despite the decidedly granular or ground-glass microscopic appearance of most *A. bovis* strains, the typical microcolony of this species is generally referred to as "smooth." *A. odontolyticus* also produces smooth microcolonies which closely resemble those of *A. bovis* (Figures 6-17 and 6-18), and, like *A. bovis,* strains of *A. odontolyticus* which produce filamentous microcolonies have been isolated. These strains have medium to long radiating filaments with or without a dense, tangled center.

*A. israelii, A. naeslundii* and *A. viscosus* all produce filamentous microcolonies, but there are some distinct differences between the species. *A. israelii* can produce several types of microcolonies, the most common and most characteristic being the "spider colony." This colony consists of branching filaments which often appear to radiate from a single point (Figures 6-19 and 6-20). The filaments are moderately long and slender with branches arising at an acute angle (Figure 6-19). Variations are seen, including: 1) tiny colonies with only one or two branching filaments (Figure 6-21), 2) somewhat larger colonies with many filaments which cross at the center (Figure 6-21) but without the tangled central mass typical of *A. naeslundii* and 3) small colonies with very short extended filaments which have an angular type of branching.

Some *A. israelii* strains produce small, compact microcolonies composed of short filaments forming a dense center with very short projecting filaments (Figure 6-22). As such colonies increase in size (48-72 hr.), they show little evidence of branching filaments but form larger, granular, compact colonies with a jagged edge and perhaps an occasional projecting filament.

The typical microcolony of *A. naeslundii* has a dense center composed of a mass of diphtheroidal cells or filaments surrounded by long, branched filaments projecting in all directions (Figures 6-23 and 6-24). Coleman et al. (78) and Howell et al. (213) report that very early microcolonies of some strains resemble the spider colonies of *A. israelii.* These colonies rapidly assumed the more typical appearance of a centered, filamentous colony. In our experience, some strains of *A. naeslundii* produce colonies which have a dense center with many very short projecting filaments.

The typical microcolony produced by hamster (209) and human (146) isolates of *A. viscosus* resembles the densely centered, filamentous colony of *A. naeslundii.* The microcolonies have a dense center composed of diphtheroidal rods with medium length projecting filaments which have angular branching (Figures 6-25 and 6-28). In some human strains, the filamentous fringe is very short (Figure 6-27) or filaments are seen on only one side of the colony. Howell (209) reported that very young, ± 12 hr., microcolonies of fresh hamster isolates were filamentous without a center and resembled *A. israelii.* These microcolonies rapidly developed into the more characteristic form with a heaped center and a filamentous fringe.

Howell (209) reported on hamster strains that dissociated during serial transfer from filamentous to smooth microcolonies due to the gradual loss of the filamentous fringe. Human isolates also tend to produce smooth microcolonies after laboratory cultivation. These colonies are circular with a highly granular surface and an entire or irregular edge (Figure 6-26). In strains which

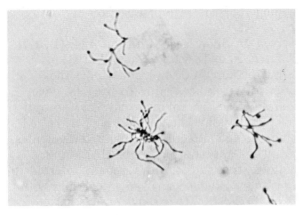

**Figure 6-19.** *A. israelii*, microcolony, filamentous "spider" colony, BHIA, 24 hr., X330.

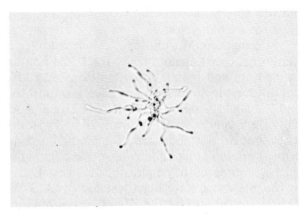

**Figure 6-20.** *A. israelii*, microcolony, filamentous "spider" colony, BHIA, 24 hr., X410. (From Slack and Gerencser, IJSB *20:*259-268, 1970.)

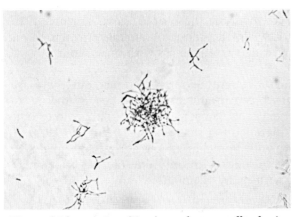

**Figure 6-21.** *A. israelii*, microcolony, small colonies with one or two filaments; larger colony resembling *A. naeslundii*. BHIA, 24 hr., X330.

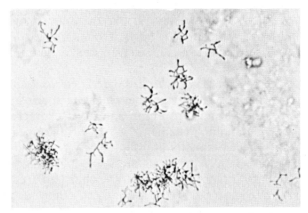

**Figure 6-22.** *A. israelii*, microcolonies, small colonies with short filaments, angular branching, BHIA, 24 hr., X410. (From Slack et al. J. Bacteriol. *97:*873-884, 1969.)

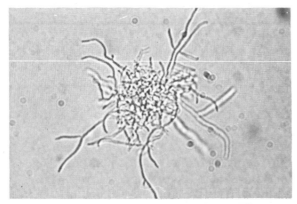

**Figure 6-23.** *A. naeslundii*, microcolony. Filamentous colony. Filamentous colony with distinct center, BHIA, 24 hr., X720.

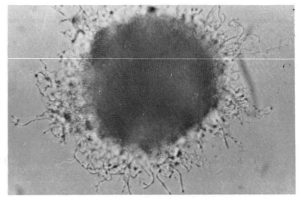

**Figure 6-24.** *A. naeslundii*, microcolony, dense center with filamentous fringe. BHIA, 48 hr., X530. (From Slack, Bergey's Manual, 1974; used with permission of the Williams & Wilkins Co., Baltimore.)

produce predominantly smooth microcolonies, a few filamentous colonies may be seen especially if very young cultures are examined.

26  Genus *Actinomyces*

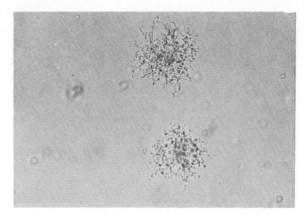

**Figure 6-25.** *A. viscosus,* microcolony, dense center with filamentous edge. BHIA, 24 hr., X330

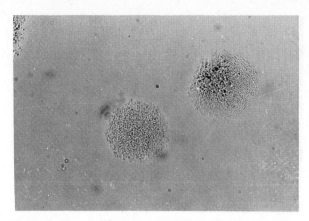

**Figure 6-26.** *A. viscosus,* microcolony, granular colony with no filaments. BHIA, 24 hr., X330.

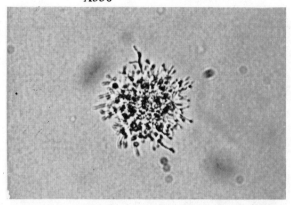

**Figure 6-27.** *A. viscosus,* microcolony, dense center with short filaments at edge and occasional longer filament. BHIA, 24 hr., X720.

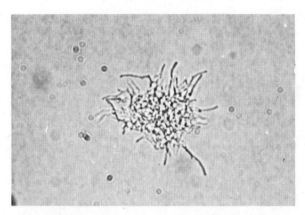

**Figure 6-28.** *A. viscosus,* microcolony, X720.

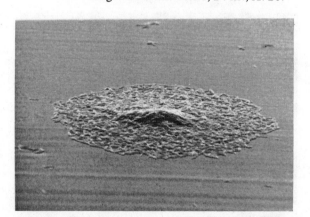

**Figure 6-29.\*** Scanning electron micrograph (SEM) of *A. bovis* microcolony, 80° tilt, 20 KV, X1000.

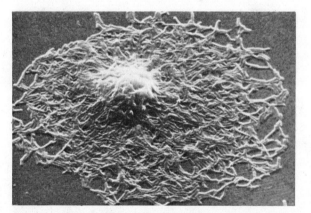

**Figure 6-30.** SEM micrograph of *A. bovis* microcolony, 45° tilt, 20 KV, X2000.

Georg et al. (140) found similar microcolony morphology in *A. viscosus* isolates from animals other than the hamster. Microcolonies were either filamentous or flat, granular and entire with dense centers, and this change in microcolony type was reflected in cell and mature-colony morphology. Most isolates produced a mixture of the two types of microcolonies. Jordan and Howell (229) found that aerobic incubation and iron depletion favored the development of

\*The scanning electron micrographs in Figures 6-29-6-35 were prepared by Phillip Allender, Division of Infectious Diseases, WVU.

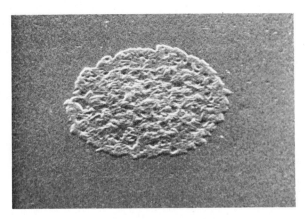

**Figure 6-31.** SEM micrograph of *A. odontolyticus* microcolony, 45° tilt, 20 KV, X2000.

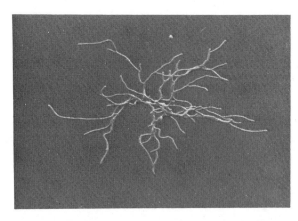

**Figure 6-32.** Scanning electron micrograph (SEM) of *A. israelii* microcolony, 45° tilt, 20 KV, X1000.

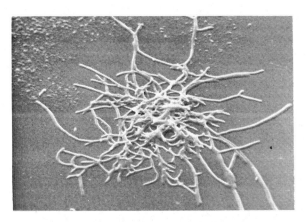

**Figure 6-33.** SEM micrograph of *A. naeslundii* microcolony, 45° tilt, 20 KV, X1000.

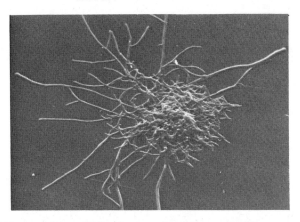

**Figure 6-34.** SEM micrograph of *A. naeslundii* microcolony, 45° tilt, 20 KV, X1000.

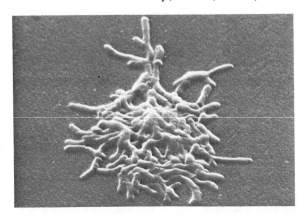

**Figure 6-35.** SEM micrograph of *A. viscosus* microcolony, 45° tilt, 20 KV, X5000.

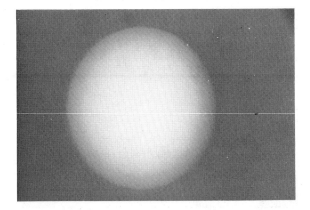

**Figure 6-36.** *A. bovis,* smooth, low convex, entire edge. BHIA, 7 days, X20.

filamentous microcolonies by hamster strains of *A. viscosus* while incubation under $N_2:CO_2$ (95%:5%) favored the smooth type. Studies of the influence of environmental factors on the microcolony type of human and other animal isolates have not been done.

The different types of microcolonies formed by *Actinomyces* are shown using the scanning electron microscope in Figures 6-29 through 6-35. The colonies were grown on strips of dialysis tubing on the surface of BHIA plates according to the technique of Afrikan et al. (5). The compact nonfilamentous colonies of *A. bovis* and *A. odontolyticus* are a sharp contrast to the

Figure 6-37. *A. bovis,* rough, center depression, knobby surface, irregular edge. BHIA, 7 days, X25.6.

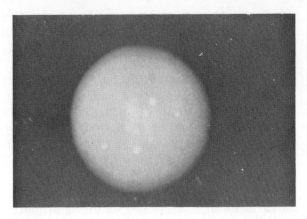

Figure 6-38. *A. odontolyticus,* smooth colony with ground-glass surface. BHIA, 7 days, X12.8.

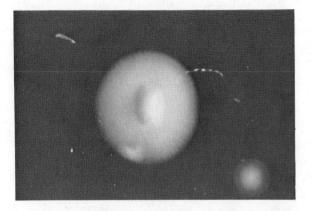

Figure 6-39. *A. odontolyticus,* flat with peaked center. BHIA, 7 days, X10.

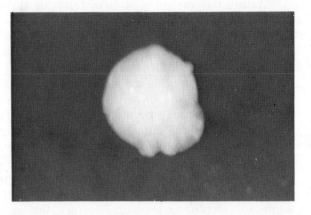

Figure 6-40. *A. odontolyticus,* smooth, umbonate. BHIA, 7 days, X12.8.

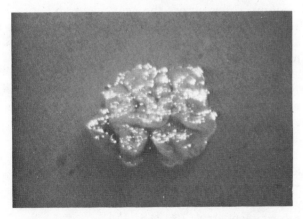

Figure 6-41. *A. odontolyticus,* rough bread-crumb colony. Blood agar, 7 days, X40.

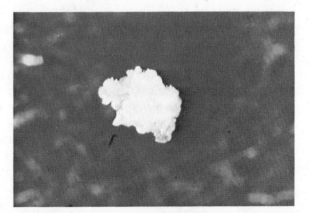

Figure 6-42. *A. israelii,* bread-crumb or molar-tooth colony, BHIA, 7 days, X25.6.

other species. The heaped-up cells in the center of the *A. bovis* colonies (Figures 6-29 and 6-30) probably account for the optically dark center seen at lower magnification. Long, well-branched filaments are seen in both *A. israelii* and *A. naeslundii* (Figures 6-32, 6-33, 6-34). The *A. naeslundii* colony shows more central growth. The *A. viscosus* colony (Figure 6-35) shows shorter filaments and a more heavily centered colony.

**Mature-Colony Morphology.** Mature colonies develop after 7-14 days on BHIA. In most instances, the colony morphology on blood agar is the same as that on BHIA. *A. bovis* colonies are 0.5 to 1.0 mm in diameter, circular, with a smooth or finely granular surface, convex, entire edge, white and usually soft (Figure 6-36). Strains which produce filamentous microcolonies also produce rough mature colonies resembling the molar-tooth colonies which will be described in detail for *A. israelii.*

*A. odontolyticus* produces a smooth mature colony which shows more variation than that of *A. bovis.* Mature colonies (Figures 6-38, 6-39, 6-40) are 1-2 mm in size, circular to irregular in shape with an entire to irregular edge. The colonies vary from low convex (Figure 6-38) to umbonate (Figure 6-40) and are smooth to finely granular, white, opaque and soft. Rough strains (Figure 6-41) which produce filamentous microcolonies produce mature colonies resembling those of *A. israelii. A. odontolyticus* is the only *Actinomyces* species which has a distinctive appearance on blood agar. The older colonies develop a characteristic deep, red color. This pigment may develop during anaerobic incubation or it may only be formed after the cultures are left standing in air at room temperature following primary anaerobic incubation. The pigment may appear in as little as 48 hrs., but it usually requires 5-10 days to develop. All of the strains studied by Batty (17) developed red pigment, but three of twenty-eight strains studied at CDC have failed to form pigment. In other respects, mature colonies on blood agar resemble those on BHIA.

Mature colonies of *A. israelii* range in appearance from convex and smooth to heaped, highly convoluted, rough colonies (48, 213, 422, 455) and usually approximate 1 mm but range from 0.5-2.0 mm in diameter. Colonies like those shown in Figures 6-42, 6-43 and 6-44 are the most frequently seen and are considered "typical, mature *A. israelii*" colonies. They are all rough with or without a central depression and have been described as: molar-tooth, bread-crumb or raspberrylike. The colonies are circular to irregular in shape with undulate, lobate or erose edges. The colony elevation from the agar surface varies from convex to pulvinate to umbonate. Some of these colonies may be more fittingly and simply described as heaped. The surface texture of the colony is often granular with a ground-glass appearance. Certain mature colonies are smooth (Figures 6-45, 6-46, 6-47, 6-48) and are circular, entire, opaque, cream, gray-white or white and soft. They may show a textured surface (Figures 6-47 and 6-48) due to differences in opacity, have concentric rings around an umbonate center or have shallow central depressions (Figure 6-48). The color of the opaque colonies under reflected light varies from white to grayish to creamy white. The colonies are usually shiny, but granular colonies may be matte in appearance. *A. israelii* colonies are usually hard and adhere to the medium. However, they may be soft and friable, and certainly roughness does not always correlate with hardness as even bread-crumb colonies may be soft. In many instances, the whole colony slides across the agar when an attempt is made to pick the colony.

In isolating *A. israelii* from dental calculus, we noted a rough colony type not previously observed. These colonies were very conical with a highly convoluted surface covered with a fuzz which seemed to be composed of short aerial hyphae (Figure 6-49). This confirms the report of such colonies by Erikson (115). The aerial hyphae, which were very numerous, consisted of gram-positive filaments and could be scraped from the surface with an inoculating loop. Two strains producing such colonies have now been isolated from different calculus specimens. They are biochemically and serologically typical *A. israelii* and have maintained the ability to produce aerial hyphae for two years.

The mature colonies of *A. naeslundii* are most commonly smooth, being 1-2 mm in diameter, circular, low convex to umbonate and entire (Figures 6-50, 6-51, 6-52). Certain strains will produce colonies with degrees of roughness (Figure 6-53) which vary from a simple granular surface to a colony which approaches the bread-crumb or raspberry (78, 213, 456).

Five-to-seven day mature colonies of hamster isolates (209) of *A. viscosus* grown under $N_2:CO_2$ (95%:5%) were approximately 0.5 mm in diameter, circular to irregular, heaped with an eccentric center depression, and were very viscous and soft in consistency. Variations occurred including colonies which were umbonate and ones with a raised, star-shaped center. When grown aerobically with 5% $CO_2$, the colonies were 0.5 to 4-5 mm, circular, entire, raised, granular, white and frosty in appearance. A few colonies were wrinkled or showed short, scattered aerial hyphae. Strains which had dissociated to the diphtheroidal form were circular, entire, low convex, smooth to finely granular, creamy white and viscous. The description of the hamster colonies grown

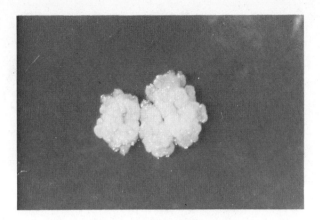

**Figure 6-43.** *A. israelii*, raspberrylike rough colony, heaped, BHIA, 7 days, X16.

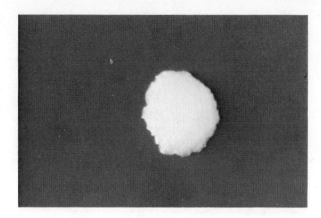

**Figure 6-44.** *A. israelii*, raised, uneven edge with relatively smooth surface, BHIA, 7 days, X40.

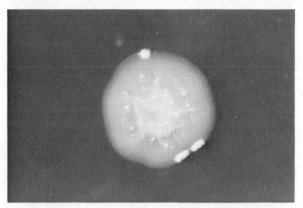

**Figure 6-45.** *A. israelii*, rough center with depression, BHIA, 7 days, X25.6.

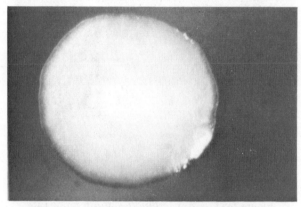

**Figure 6-46.** *A. israelii*, smooth umbonate, BHIA, 7 days, X80.

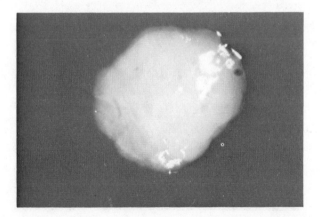

**Figure 6-47.** *A. israelii*, smooth convex, textured surface, BHIA, 7 days, X80.

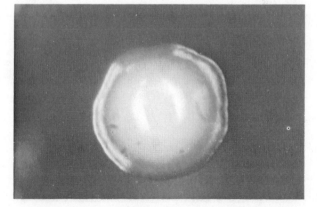

**Figure 6-48.** *A. israelii*, smooth, with textured surface and shallow center depression, BHIA, 7 days, X80.

aerobically with $CO_2$ fits our observations of human strains (Figures 6-54–6-58) quite well except that we have not observed aerial hyphae. Colonies of the human isolates varied mainly in elevation and surface topography. The variations observed include the presence of radial or concentric striations (Figures 6-57 and 6-58) and shallow central depressions. The conditions of incubation, aerobic or anaerobic, had little affect on the colony type with the human isolates.

**Figure 6-49.** *A. israelii,* rough colony with aerial hyphae, BHIA, 7 days, X25.6.

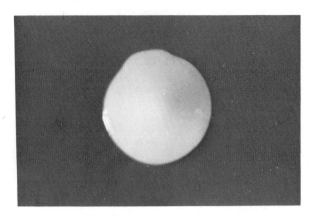

**Figure 6-50.** *A. naeslundii,* smooth convex, BHIA, 7 days, X8.

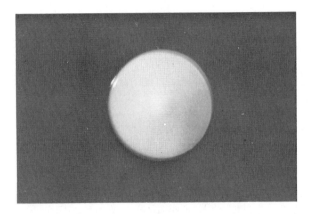

**Figure 6-51.** *A. naeslundii,* smooth umbonate, BHIA, 7 days, X6.4.

**Figure 6-52.** *A. naeslundii,* smooth convoluted surface with umbonate center and erose margin, BHIA, 7 days, X20.

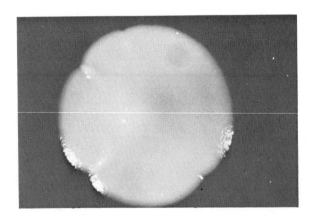

**Figure 6-53.** *A. naeslundii,* granular with scalloped edge, umbonate, BHIA, 7 days, X10.

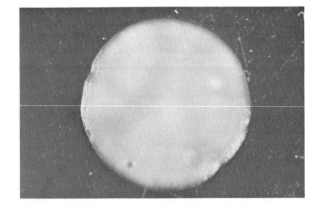

**Figure 6-54.** *A viscosus,* smooth, slightly umbonate, textured surface, BHIA, 7 days, X12.8.

Georg et al. (140) described both smooth and rough mature colonies for strains of *A. viscosus* isolated from dogs, pigs and goats. The rough colonies were typical molar-tooth types resembling those of *A. israelii.*

The appearance of colonies of *A. israelii, A. naeslundii, A. bovis* and *A. viscosus* on blood agar is essentially the same as on BHIA. None of these species is hemolytic. As previously indicated, *A. odontolyticus* produces a characteristic deep-red colony on blood agar, and young colonies may be either α or β hemolytic.

**Figure 6-55.** *A. viscosus,* smooth, low convex, BHIA, 14 days, X40.

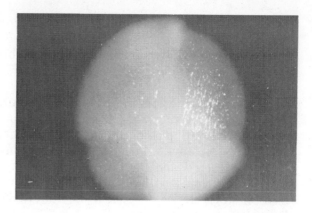

**Figure 6-56.** *A. viscosus,* convex, wedge-shaped sectors, "frosty" surface. BHIA, 7 days, X20.

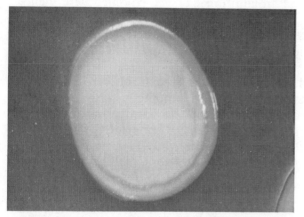

**Figure 6-57.** *A. viscosus,* smooth, oval, concentric ring, BHIA, 14 days, X40.

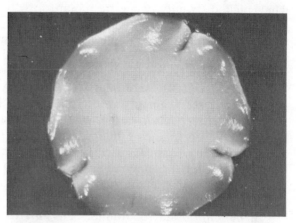

**Figure 6-58.** *A. viscosus,* smooth, convex with radial indentations, BHIA, 14 days, X40.

**Figure 6-59.** *A. suis,* heaped, bread-crumb colony, BHIA, 7 days, X12.8.

**Figure 6-60.** *A. suis,* smooth, convex, with erose edge, BHIA, 7 days, X20.

The swine isolates described by Franke (128) produce colonies similar to those of *A. israelii.* One of the strains we examined produced very knobby, bread-crumb type colonies (Figure 6-59) while the second strain produced a convex colony with a smooth surface in the center and a deeply indented rough edge (Figure 6-60).

**Occurrence of Colony Types.** The frequency with which the various colony types occur affects the usefulness of morphology for isolation and identification. We have no data on the frequency with which smooth and rough *A. bovis* types occur, but apparently the great majority of isolates are smooth. Pine et al. (370) found filamentous colony types among isolates from one of four cases of bovine actinomycosis. Some information concerning the occurrence of colony types of the other four species can be obtained from our studies on the isolation of *Actinomyces* from dental calculus.

*A. israelii* **Microcolonies.** Thirty-eight colony isolates of *A. israelii* from eleven different calculus specimens were examined for microcolony type on initial isolation, that is, on the first plate streaked after picking the colony. Twenty-three of the thirty-eight produced typical filamentous microcolonies; nine produced very tiny, compact colonies; three produced colonies suggestive of *A. naeslundii*; two had a heaped center with short projecting filaments; and two were granular and heaped without projecting filaments. These results supported an earlier estimate based on stock strains (424) that only about two-thirds of *A. israelii* strains produce typical filamentous microcolonies.

*A. israelii* **Macrocolonies.** The original isolation plates were incubated 5-7 days; the macrocolony types were then described and picked. Sixty-four colonies which were later identified as *A. israelii* were picked from sixteen different specimens. Nine of the colonies were still filamentous when picked; twenty-two were molar-tooth types; and an additional twelve colonies were highly granular, convex or umbonate. Five colonies were rough with aerial hyphae. Twelve colonies were smooth, convex to slightly umbonate, and three were smooth with a textured surface. It should be noted that, in several cases, both rough and smooth colonies were picked from the same set of isolation plates.

*A. naeslundii* **Microcolonies.** The majority of *A. naeslundii* colony isolates (20/34) produced typical, centered filamentous microcolonies. Two strains produced filamentous colonies without a distinct center (*israelii*-type), while the remaining isolates produced less filamentous microcolonies. Seven isolates produced coarsely granular colonies which were compact and had very short projecting filaments, and five strains produced granular colonies which had no filaments or only an occasional filament. In the case of *A. naeslundii*, which grows very rapidly, the compact nature of some of the colonies may be due to age. It is often necessary to observe plates at less than 24 hrs. if truly filamentous colonies are to be seen.

*A. naeslundii* **Macrocolonies.** The mature colonies of the thirty-four dental isolates showed all the variations described for *A. naeslundii* in the section above. The most common colony type (15/34) was the smooth type shown in Figure 6-50. These colonies occasionally had undulate or lobate edges but were otherwise the same as the smooth colonies. Five colonies had radial grooves at the edge, and seven were rough with varying degrees of surface irregularity.

*A. viscosus* **Microcolonies.** Four of twenty-one colony isolates of *A. viscosus* had centered, filamentous microcolonies; seven had more compact microcolonies with a short filamentous fringe; and six had dense, irregular colonies with only an occasional projecting filament. Four strains had entire, granular microcolonies which were not filamentous. As in the case of *A. naeslundii*, rapid growth and subsequent rounding-up of the microcolony may explain why more strains did not have highly filamentous microcolonies.

*A. viscosus* **Macrocolonies.** Forty colonies of *A. viscosus* were described on the original isolation plates. Three of these still had a filamentous edge, and six had rough colonies resembling molar-tooth colonies. The remainder were essentially smooth but showed variation in elevation which ranged from smooth to umbonate (twenty-one strains), marginal rings or center depressions (seven strains) or a ground-glass surface (three strains).

*A. odontolyticus* **Microcolonies and Macrocolonies.** This species was the least variable. All the calculus isolates had typical smooth microcolonies, although we have isolated a filamentous strain from lacrimal canaliculitis (145). Fifteen of the twenty-one macrocolonies isolated on blood agar showed some degree of red pigment when picked. Most isolates (17/21) were smooth, although they varied from convex to umbonate and from entire to crenate edges. Three colonies were granular or heaped on the isolation plates.

**Conclusion.** While colony types are a useful and important characteristic, they cannot be relied upon exclusively either for selecting possible *Actinomyces* on isolation plates or for identifying isolates as to species.

## B. ULTRASTRUCTURE*

The few ultrastructural studies done on the genus *Actinomyces* include the work of Edwards and Gordon (108), Overman and Pine (348), Duda and Slack (101) and Girard and Jacius (159).

Overman and Pine (348) directed their study to what they called "cytoplasmic figures" (mesosomes) in hopes of using the occurrence and complexity of these structures for taxonomic purposes. They found that *A. bovis* had no mesosomes or only short, straight, membranelike structures, while *A. israelii* and *A. naeslundii* contained simple, coiled structures which appeared to be connected to the cytoplasmic membrane. In contrast, *A. propionica* contained compact, highly coiled mesosomes which had no apparent contact with the cytoplasmic membrane. They also noted distinct differences in cell-wall thickness. The most striking difference was between *A. israelii* (29 nm) and *A. bovis* (10 nm). *A. naeslundii* (20 nm) resembled *A. israelii*, while the *A. propionica* wall was intermediate (11.4 nm). These authors comment on the presence of a debrislike material on the cell wall of cells fixed by the Kellenberger method which was not present on walls fixed by the method of Chapman and Hillier.

Electron microscope studies of *Actinomyces* done in this laboratory, some of which were reported by Duda and Slack (101), will be described in some detail. Duda and Slack (101) studied all five species to not only confirm differences in the species which might be used for taxonomic purposes but to further study the ultrastructural details of *Actinomyces* and their mode of growth and cell division. They were unable to confirm the differences in cytoplasmic structures reported by Overman and Pine (348) but did confirm differences in cell-wall thickness (see Table 6-1).

The organisms were grown in 300 ml of BHI broth in an atmosphere of $N_2$-$CO_2$ (95%:5%) at 37 C for 1-7 days. Cells were prepared for electron microscopy by pouring the broth culture over frozen and crushed BHI broth and harvesting by centrifugation at 4 C. The chilled cells were fixed according to the methods described by Kingsbury and Voelz (247), Kellenberger et al. (239) and Falk (119). Agar-embedded cells were dehydrated through a graded series of ethanol or acetone and embedded in Epon 812 or Vestapol W.

The five species exhibited similar cell morphologies in ultrathin sections and were in no way different from other gram-positive bacteria. Cells were seen as bacillary forms (Figure 6-61), and clusters of cells showed the typical branching morphology of the genus (Figures 6-62, 6-63, 6-64). Various bizarre cell types with a markedly thickened, often hyaline cell wall were found (Figure 6-65). These occurred most frequently in old cell populations.

The *cystoplasmic membrane* was not readily discernible and was tightly bound to the cell wall (Figure 6-64). However, it could be separated from the cell wall by plasmolysis (Figures 6-65 and 6-66), demonstrating its rigidity and intimate attachment to the cytoplasm. The cytoplasm of young, actively growing cells, which stained darkly with lead citrate, was granular and often packed with ribosomes (Figures 6-61, 6-63, 6-64). The *cytoplasm* of older cells stained

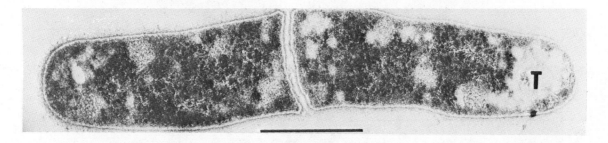

**Figure 6-61.** Four-day-old, rod-shaped cells of *A. bovis* fixed by the R-K method. Division is by septa formation. Nuclear material is sparse and distributed in spots throughout the cytoplasm. T, translucent areas. X70,000. The bars in all photographs represent 0.5 μm except where noted.

*The electron microscope studies were done by John J. Duda, WVU.

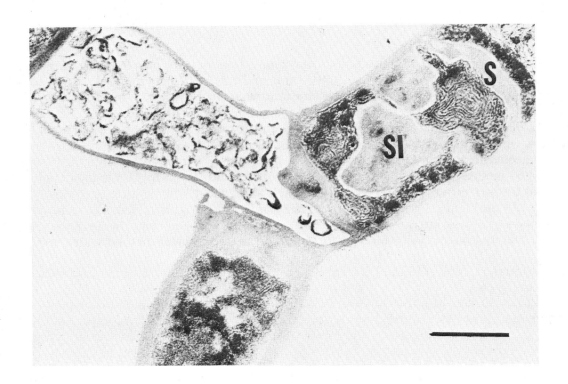

Figure 6-62.   A "Y" cell cluster of a 7-day-old culture of *A. naeslundii* fixed by the R-K method. Cells are beginning to degenerate or have degenerated.  S, completed septum.  SI, incompleted septum.  X40,000.

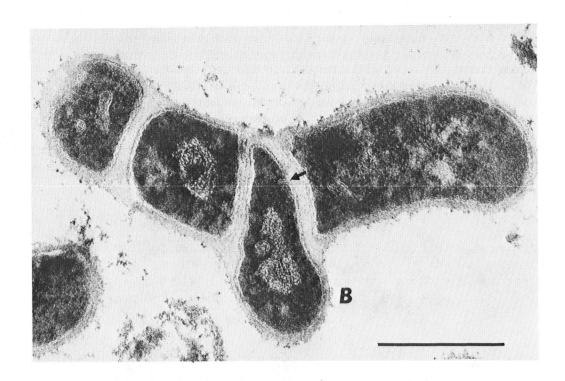

Figure 6-63.   A "Y" cell cluster of 4-day-old *A. odontolyticus* fixed with glutaraldehyde followed by R-K method. A thick cell wall with a granular material is noted.  B, budding cell.  Arrow, membrane invagination. X85,000.

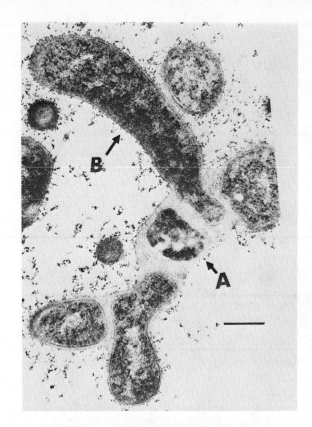

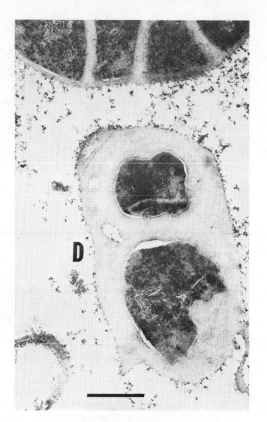

Figure 6-64. Demonstration of "Y" and "T" forms in a cluster of cells of a 4-day-old *A. odontolyticus* culture fixed by the R-K method. Note the difference in the width of cell walls between "budding" cell (B) and the adjacent cell (A). X40,000.

Figure 6-65. Bizarre cells from a 4-day-old culture of *A. odontolyticus* fixed by the R-K method. Note the extremely thick cell walls of the apparently degenerating cells (D). X36,000.

erratically, often showed autolysis, disintegration, widespread membranous growth and a lack of ribosomes (Figure 6-62).

The *nuclear apparatus* had no membrane and tended to be centrally located, roughly spherical (Figures 6-63 and 6-64) and granular or fibrillar (Figure 6-67). It stained intensely with osmium tetroxide. The granular material within the apparatus was tightly packed or loose, while the fibrils most often showed a loose arrangement (Figures 6-63, 6-64, 6-66).

*Cytoplasmic inclusions* were present at some time in all species and appeared as either electron opaque or translucent areas (Figure 6-68). In an older cell population, a tubular symmetrical type of structure was noted (Figure 6-69). This symmetry could represent different planes of sectioning through a tubule or mesosome. The nature of these structures has not been determined.

*Mesosomes* occurred in all species, but their number, size and type varied widely and were inconsistent even in a single species (Figures 6-62, 6-63, 6-66). The most common type was a tubular structure (Figure 6-63), although membranous types (Figure 6-62) were noted in older cells. Mesosomes were associated with the cell membrane at septa formation sites (Figures 6-63 and 6-70); however, there was no indication that these structures could aid in the completion of septa after cell division. After division, mesosomes were often present in both cells, although they were sometimes elaborately formed in one daughter cell and only sparsely formed in the other cell (Figure 6-66). Mesosomes were also often seen in close proximity to the nuclear apparatus (Figure 6-63).

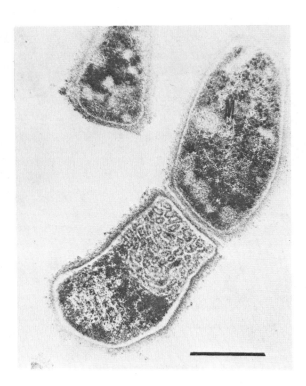

Figure 6-66.    Oblique sections through a 4-day-old cell of *A. bovis* fixed by R-K method.  Extensive infolding of the membrane in one daughter cell near the division plane.  The other daughter cell contains only one mesosome (M).  Cytoplasmic membrane has separated from cell wall.  X45,000.

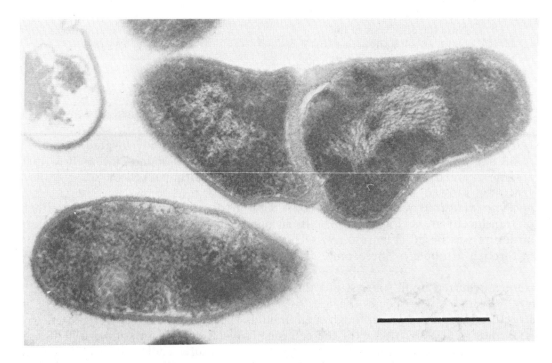

Figure 6-67.    Seven-day-old cells of *A. odontolyticus* fixed in glutaraldehyde and 2% osmium tetroxide in cacodylate buffer.  Note loose fibrillar arrangement of nucleoid apparatus in one cell and coarse granular and fibrillar arrangement in adjacent cell.  X60,000.

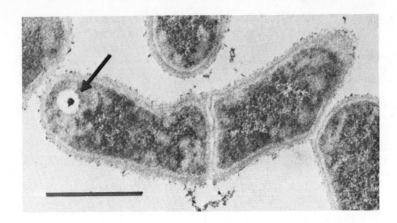

Figure 6-68.    Four-day-old cells of *A. viscosus* prefixed in glutaraldehyde followed by R-K method.  Cytoplasmic inclusion (arrow).  X50,000.

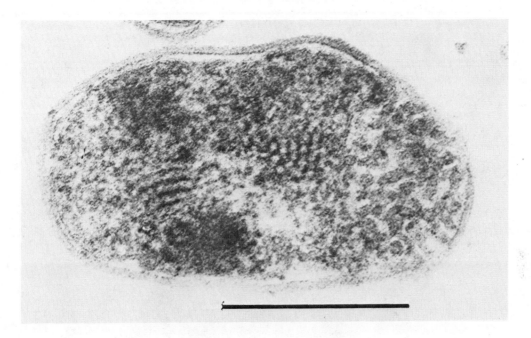

Figure 6-69.    Seven-day-old cells of *A. odontolyticus* fixed by the R-K method. Symmetrical parallel bars and tubular structures.  X100,000.

One of the most striking characteristics of the *Actinomyces* was the thick *cell wall* which often had a fuzzy or shaggy appearance (Figures 6-61, 6-63, 6-70).  The wall thickness was measured in normally growing cells of all species grown for the same length of time in the same growth medium.  Cells which appeared to have synthesized and deposited excessive amounts of cell-wall material were excluded.  The wall thicknesses of the five *Actinomyces* species are shown in Table 6-1.  All of the wall measurements are greater than those reported by Overman and Pine (348), but the relationship between *A. israelii* and *A. bovis* was similar.  The cell wall appeared to be composed of two layers, an inner, thin, darkly staining layer and an outer, amorphous, lightly staining layer (Figures 6-63, 6-64, 6-71).  This double structure was found in cells fixed by various methods, in unfixed cells and in freeze-etch preparations.

The *outer layer* with its fuzzy edge makes up the greater portion of the wall and apparently is polysaccharide in nature.  Attempts to increase the thickness of the outer wall by growing the organism in increasing concentrations of glucose and sucrose were not successful with *A. israelii*.

Table 6-1. Cell-wall thickness of five strains of
*Actinomyces*. Cells grown for 66
hours in BHI broth.

| Organism | Cell-wall thickness |
|---|---|
| *A. israelii,* ATCC 12102 | 64 nm |
| *A. naeslundii,* ATCC 12104 | 45 nm |
| *A. viscosus,* ATCC 15987 | 35 nm |
| *A. bovis,* ATCC 13683 | 31 nm |
| *A. odontolyticus,* ATCC 17929 | 30 nm |

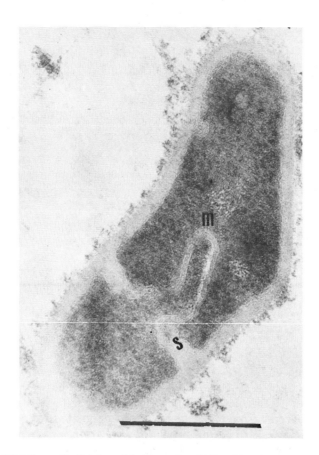

**Figure 6-70.** Dividing cell of 4-day-old *A. naeslundii* fixed by the R-K method. S, septa.
M, advancing growth of membrane. X75,000.

A silver stain specific for polysaccharides (356, 412, 453) was used to identify cell-wall constituents cytochemically. The fine grains of silver were deposited over the entire cell, but there was a symmetry to those deposited on the wall (Figure 6-73) indicating that polysaccharide is a cell-wall constituent. The *inner layer* of the cell wall is thin and more darkly staining than the outer

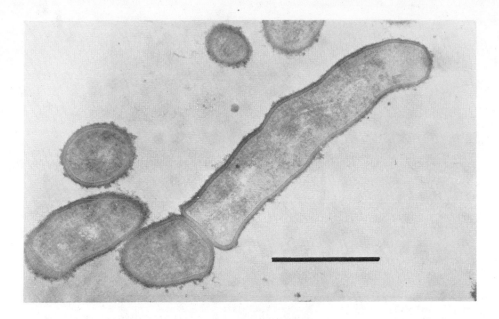

**Figure 6-71.** Four-day-old cells of *A. israelii* fixed according to the method of Luft and stained with ruthenium red. The two-part structure of the cell wall is clearly delineated. Bar equals 1.0 μm. X29,000.

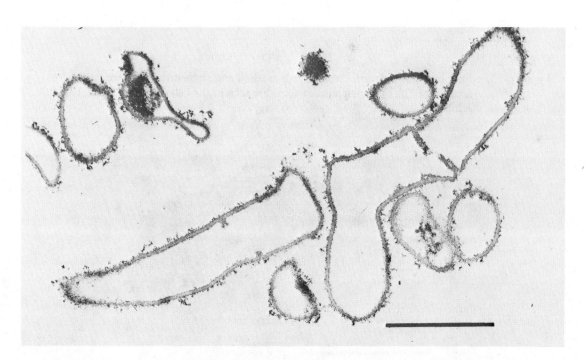

**Figure 6-72.** Purified cell walls of 7-day-old *A. israelii* obtained by extraction with alkaline ethyl alcohol and pronase according to method of Freer et al. Double-track structure of cell wall is clearly visible. Bar equals 1.0 μm. X28,000.

layer. Efforts to break down this layer to form protoplasts using lysozyme, snail gut juice and other enzymes have so far been unsuccessful. In cell-wall preparations of *A. israelii* purified by the method of Freer et al. (130), only the inner layer is retained. In these preparations, the inner layer is itself a double-track structure (Figure 6-72).

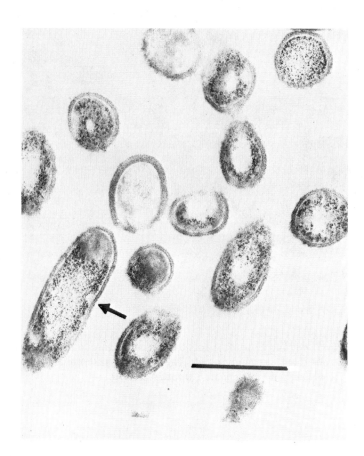

**Figure 6-73.** Four-day-old *A. israelii* fixed by R-K method and stained for polysaccharides by method of Thiery. Fine grains of silver are seen in the cell wall. Bar equals 1.0 $\mu$m. X25,000.

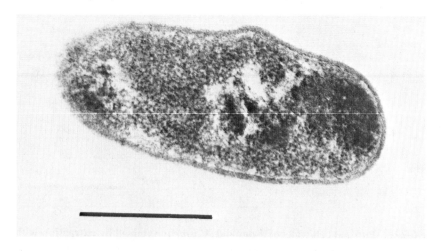

**Figure 6-74.** Seven-day-old *A. odontolyticus* fixed by R-K method. Single cell bulging laterally. X70,000.

Cell division was by binary fission (Figure 6-61); however, other interesting features were noted including the development of cell clusters which suggested that division occurred in two different directions. First, there was typical binary fission resulting in a straight row of cells

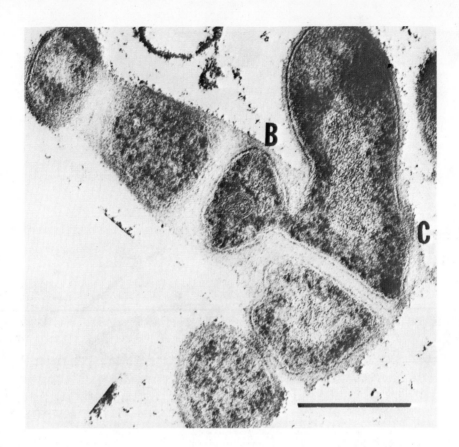

**Figure 6-75.** Cell cluster of a 4-day-old *A. odontolyticus* fixed by the R-K method. The central cell (C) has retained a cytoplasmic bridge to the "budded" cell (B). X60,000.

which had divided but did not separate (Figure 6-64). At a right angle from the row, there was another cell or row of cells which may have resulted from a different mode of division (Figures 6-64 and 6-75). Budding as a means of division was suggested by the presence of single cells which seemed to bulge laterally (Figure 6-74) and by other cells which had formed a septum but retained a bulge located adjacent to the septum (Figure 6-67). These often formed a second septum in such a way that the budding cell was bordered by two non-bulging cells (Figure 6-64). This cell is referred to as the center cell and may be identical to the laterally out-growing cells seen in Figure 6-75.

Girard and Jacius (159) studied electron micrographs of shadowed, negatively stained, thin-sectioned and freeze-etched cells of *A. viscosus* and *A. naeslundii.* They studied both 1- and 5-day-old cells and demonstrated a greater number of dead and degenerating cells in the 5-day suspensions. They question some of the morphological features described by Duda and Slack (101) indicating that the cultures may have been too old. However, the contrast between young, old and degenerate forms was pointed out, and the young cells closely resemble those of Girard and Jacius (compare Figures 6-63 and 6-77). These authors describe very interesting fibril-like structures on the outer surface of both *A. naeslundii* and *A. viscosus* (Figures 6-76 and 6-78). These fibrils have a diameter of 4-8 nm and extend out from the wall. They are most prominent in *A. viscosus.* The authors suggest that the fibrils could serve as a mechanism of attachment to the enamel surface of teeth and between the organisms themselves. They indicate that the cell wall of *A. naeslundii* is thicker than the wall of *A. viscosus*, but both show a double-layered structure (Figure 6-77). In freeze-etch preparations (Figure 6-78) the plasma membrane has two layers, the outer of which contains numerous particles while the inner is smooth with evenly dispersed domelike structures.

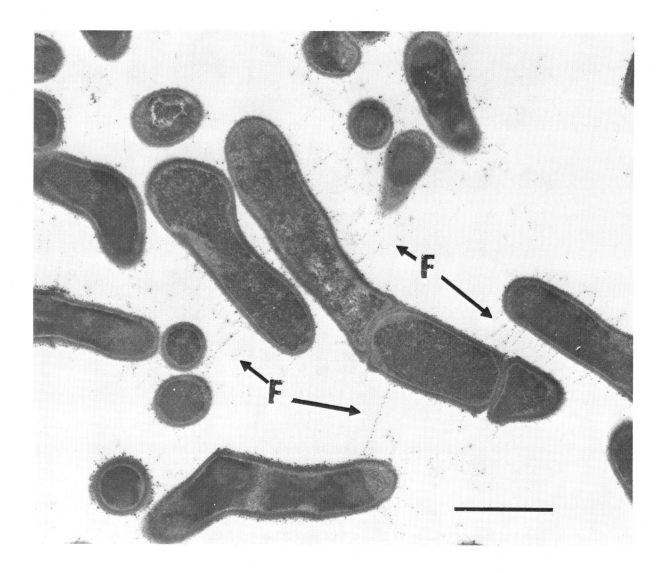

Figure 6-76.    Thin section of *A. viscosus* (24 hr.).  The "fuzzy" layer (F) is seen as stringlike fibers extending out into the medium and joining one cell to another.  X21,600.  (From Girard and Jacius, Arch. Oral Biol. *19:*71-79, 1974; used with permission of Pergamon Press, Elmsford, New York.)

## C.  OXYGEN REQUIREMENTS

The members of the genus *Actinomyces* have been referred to as obligate anaerobes, facultative anaerobes and micoaerophiles by various authors. This confusion results from the multiplicity of definitions of these terms as well as variations in methods, conditions and interpretation of tests used to determine oxygen requirements.  This problem will not be resolved until methods are developed to actually determine the gaseous requirements of a particular organism and express these in mathematical terms and thus eliminate the confusing terminology.  In the meantime, we are defining these terms for use in this monograph as follows:

**Anaerobe.**  An organism which grows well on the surface of solid medium through successive transfers when incubated in an anaerobic jar (catalyst type).  Does not grow or produces minimal growth on the surface of aerobic plates.  Does not form surface colonies in agar shake cultures.

**Facultative anaerobe.**  An organism which grows well and survives successive transfers on the surface of agar plates incubated aerobically or anaerobically.  May or may not grow on the surface of agar shake cultures.

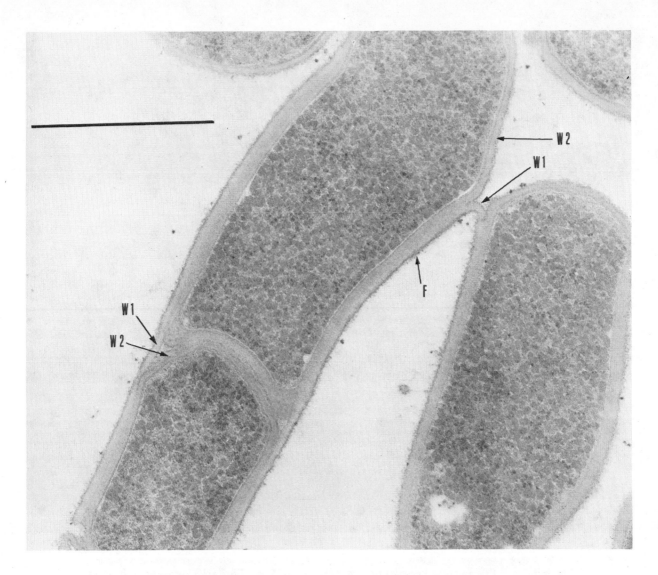

Figure 6-77.    Thin section of *A. naeslundii* (24 hr.).  The "fuzzy" layer (F) is seen as well as the layered cell wall of this organism.  The outer layer (W-1) and inner layer (W-2) are seen at the region of septation.  Bar equals 1.0 $\mu$m. X58,200. (From Girard and Jacius, Arch. Oral Biol. *19*:71-79, 1974; used with permission of Pergamon Press, Elmsford, New York.)

**Microaerophile.**  An *obligate aerobe* which grows more rapidly at partial pressures of oxygen substantially below atmospheric pressures.  This definition was used by the editors of the eighth edition of Bergey's Manual.

According to these definitions, *A. bovis* and *A. israelii* are anaerobes while *A. naeslundii, A. viscosus* and *A. odontolyticus* are facultative anaerobes.  None of the species are microaerophilic, and this term should no longer be applied to these organisms.

Although *A. bovis* is considered an anaerobe, it is generally more aerotolerant than *A. israelii* (115, 431).  When a large inoculum (O.D.=0.5) is used (370), most strains will grow aerobically with added $CO_2$, but maximum growth only occurs anaerobically and small inocula will only grow anaerobically.

*A. israelii* grows best under anaerobic conditions (115, 143, 202, 217, 277, 338, 390, 423, 455).  There is strain variation in $O_2$ tolerance with some showing only minimal growth and some (e.g., strain ATCC 10049) resembling *A. naeslundii* in the amount of aerobic growth (362). Howell et al. (213), using liquid casitone-starch-glucose medium, demonstrated that approximately

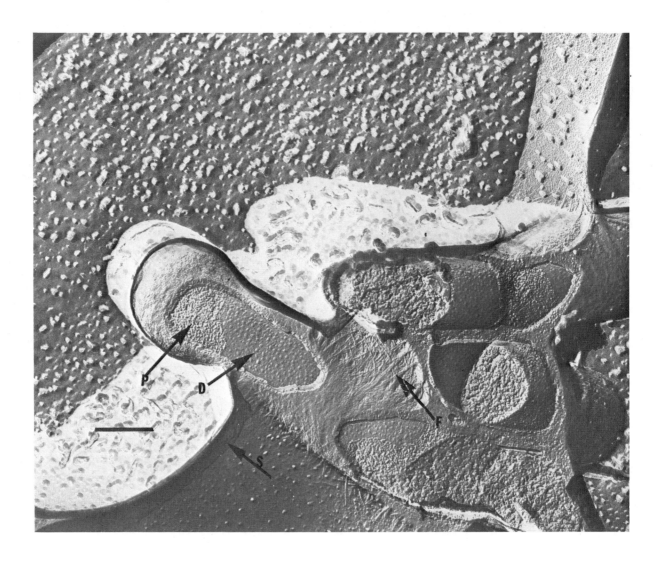

**Figure 6-78.** Freeze-etched preparation of *A. viscosus.* **Many fibrils (F) are seen on the outer surface of the cell wall. Just below the cell wall is a layer composed of numerous particles (7-8 nm in diameter) and beneath this layer is a relatively smooth surface covered by evenly dispersed "domes" (D) or hemispheres about 14-17 nm in diameter. The direction of the shadow is indicated by arrows. Bar equals 1.0 μm. X56,000. (From Girard and Jacius, Arch. Oral Biol.** *19:***71-79, 1974; used with permission of Pergamon Press, Elmsford, New York.)**

75% of *A. israelii* strains produced 2-3+ growth when grown aerobically with $CO_2$ using a large inoculum (O.D.=0.5). When this inoculum was diluted 1:10,000, none of the strains grew well aerobically, but they did grow well anaerobically with $CO_2$. When a similar method was used with BHIA slants and seals giving various atmospheres (see Chapter 14), Brock and Georg (48) and Slack et al. (424) found that seventy-one of ninety strains of *A. israelii* gave good growth (2+) aerobically with added $CO_2$. These strains were not tested for their ability to initiate growth from small inocula or to survive repeated transfers under aerobic conditions. Others report that *A. israelii* frequently failed to grow aerobically from small inocula and died out on repeated aerobic transfer (369, 391). When *A. israelii* was tested in agar deeps (Chapter 14), none of the strains grew on the surface. A few strains showed a zone of maximum growth as little as 2-3 mm below the surface, but, for the majority, maximum growth began 7-8 mm below the surface.

*A. naeslundii* was first recognized as a distinct entity on the basis of its ability to grow well aerobically (332, 456). Some of the variation in earlier descriptions of the oxygen requirements of *A. israelii* is probably due to the inclusion of unrecognized strains of *A. naeslundii* in the cultures studied. Howell et al. (213) found that *A. naeslundii* was much more oxygen tolerant than *A. israelii*, but the ability of both species to grow aerobically with added $CO_2$ was highly dependent on inoculum size and with some strains on the medium constituents. All *A. naeslundii* strains grew well aerobically with $CO_2$ when a heavy inoculum (O.D.=0.5) was used, and 64% continued to grow well aerobically with a smaller inoculum. But with the small inoculum, *A. naeslundii* strains grew best anaerobically with $CO_2$. Coleman et al. (78), using the agar slant method, found that four out of twelve strains of *A. naeslundii* grew equally well aerobically and anaerobically with no obvious requirements for $CO_2$. Seven strains grew best with $CO_2$ (aerobically or anaerobically), while one strain grew only under anaerobic conditions. In our laboratory, *A. naeslundii* strains usually grow equally well aerobically and anaerobically and may or may not show a preference for $CO_2$. When forty-six strains were tested by the agar deep method, only four grew on the surface; thirty-six showed maximum growth at 1-4 mm below the surface; and six showed maximum growth at 4-6 mm below the surface.

Howell (209), in his original report on the hamster organism, found that *A. viscosus* was essentially aerobic and required $CO_2$ for good growth on most media. While undoubtedly the most aerobic of the *Actinomyces, A. viscosus* is more accurately described as a facultative anaerobe. In our hands, *A. viscosus* grows best aerobically with $CO_2$, but good growth is obtained aerobically without $CO_2$ and anaerobically with $CO_2$. Growth is generally more rapid in the presence of $CO_2$, but after seven days slants with and without $CO_2$ have comparable amounts of growth. In agar deeps, most strains had a zone of maximum growth just under the surface with little or no growth on the surface, but with five of twenty-seven strains the zone of maximum growth extended to the surface. Dog, pig and goat strains of *A. viscosus* (140) grew better aerobically than anaerobically, but in all cases growth was improved by the addition of $CO_2$.

Batty (17) reported that all strains of *A. odontolyticus* grew equally well under aerobic and anaerobic conditions on blood agar plates. The presence of blood in the medium apparently affects the requirement for oxygen since the sixteen strains tested at CDC (9) grew poorly or not at all on BHIA slants incubated aerobically. They usually grew well (13/16) aerobically with added $CO_2$, but optimum growth was obtained anaerobically with or without $CO_2$. In this laboratory, twelve dental isolates of *A. odontolyticus* which were tested in agar deeps gave varied areas of maximum growth. Four strains showed maximum growth beginning at more than 10 mm below the surface, three beginning at 2-5 mm and four beginning at 1 mm. Only one strain gave surface growth.

While a definite requirement for $CO_2$ may be difficult to demonstrate under the routine test conditions, Pine (362) has shown that *A. israelii, A. bovis, A. naeslundii* and *A. viscosus* require $CO_2$ for anaerobic growth. All except *A. naeslundii* also require $CO_2$ for aerobic growth. *A. odontolyticus* does not seem to require $CO_2$ for aerobic or anaerobic growth.

## D. BIOCHEMICAL REACTIONS

**Introduction.** The biochemical reactions of the five *Actinomyces* species are shown in Tables 6-2 through 6-8. The tables show combined results from the literature and those obtained in our laboratory. Data for all tests are not available for all species or for all strains of any one species. The tables have been designed to include as many biochemical tests as possible and to show the range of strain variation within a species. Wherever the results from only one strain are recorded, this data is for the type, neotype or suggested neotype strain studied by members of the Subgroup. The media and methods used in our laboratory are given in Chapter 14. Where different methods were used in the literature, these are indicated to aid in distinguishing between strain variation and variation due to methodology. It can be shown that some tests, i.e., fermentation tests and urease tests, are markedly influenced by the basal medium. For example, *A. naeslundii* produces acid from glycerol when the basal medium is Thioglycollate Fermentation Base (TFB) but not when Actinomyces Fermentation Broth (AFB) is used even though good growth is obtained in both media.

Little is known about the effect of differing degrees of anaerobiosis on biochemical tests. Anaerobiosis has been obtained by a variety of methods including the use of deep agar stabs, thioglycollate-containing media, sealing of individual tubes with alkaline pyrogallol seals and incubation in anaerobe jars of various types. The VPI anaerobe laboratory (201) has studied *Actinomyces* and *Arachnia* using stringent anaerobic conditions. The five *Actinomyces* species do not have the same optimal oxygen requirements—a factor which has not usually been considered in doing biochemical tests. Melville (315) compared biochemical tests done both aerobically and anaerobically on strains capable of growing under both conditions and found little difference in his overall similarity values. Nevertheless, this point needs to be investigated both in comparison of strains of a species and in comparison of different species.

At this point, comments applicable to the genus as a whole seem appropriate. All species of *Actinomyces* are indole negative, nonproteolytic and, with the exception of *A. viscosus,* catalase negative. Some strains of *A. israelii* and about 50% of the strains of *A. humiferus* will produce a zone of hydrolysis on gelatin agar plates flooded with mercuric chloride, but they do not liquefy gelatin in standard tube tests. Reports of *Actinomyces* cultures which liquefy gelatin or produce indole are almost certainly due to the use of a mixed culture or to confusion of *P. acnes* with *Actinomyces.* Only *A. viscosus* is catalase positive in that it causes the evolution of gas bubbles from hydrogen peroxide. However, Hammond and Peindl (184) have demonstrated the presence of catalase enzymes in *A. naeslundii* by disc gel electrophoresis.

All species of *Actinomyces* ferment glucose. The end products of the fermentation are acetic, formic (tr), lactic and succinic acids. Fermentation of other carbohydrates is both species and strain variable. Hydrogen sulfide production depends on the medium and on the method of detection. Most strains of all species are hydrogen sulfide negative when heart infusion agar or BHIA are used with lead acetate strips. However, when triple sugar iron agar was used, all strains tested produced sufficient $H_2S$ to blacken lead acetate strips but did not change the indicator in the medium. There are varying reports on the production of acetyl-methyl carbinol by *Actinomyces.* In our hands, the Voges-Proskauer test has been difficult to interpret with poor reproducibility so that we no longer use this test routinely.

A discussion of each species follows, giving additional information and pointing out particular problems. For further details consult the tables.

*Actinomyces bovis* (Table 6-2). *A. bovis* is less active biochemically than the other species. Its most distinctive reaction is the rapid hydrolysis of starch with the production of a wide zone of hydrolysis in 24-48 hrs. Esculin is hydrolyzed, but nitrate and urease tests are negative. Glucose, fructose, lactose and starch are fermented by 90% of the strains. Maltose and sucrose are usually fermented, but mannitol, raffinose and rhamnose are usually negative. The reactions on glycerol, xylose and salicin are variable and appear to be somewhat medium dependent.

Even though *A. bovis* is the oldest known and the type species of the genus, there is only a meager amount of information available on verified strains. We were able to collect information on only thirty-one strains and even here only a few tests include all the strains. The data in the table can be supplemented by the following additional reports in the literature. Erikson (115) studied five strains with only a limited number of carbohydrates, but these agree with those listed in the table. Georg (136) supplements our data by indicating that inositol is usually fermented and that results with trehalose are variable while adonitol, inulin and sorbitol are not fermented. The VPI laboratory (201) studied two strains extensively under strictly anaerobic conditions. The two strains were not identical, but the results were in general agreement with those listed in the table. Under these conditions, weak reactions were found in a number of sugars which are negative in our table.

*Actinomyces odontolyticus* (Table 6-3). The results given in the table are based on unpublished studies from our laboratory and on those of Georg (137). This species reduces nitrate and about 75% of the strains hydrolyze esculin. Starch is not hydrolyzed, although it is usually fermented. Urease is negative on Thioglycollate Urease Broth and on Christensen's Urea agar. Glucose, starch, sucrose and usually ribose are fermented but not adonitol, dulcitol, mannitol or

Table 6-2. Biochemical reactions of *Actinomyces bovis*[a].

| Test | Strains Positive[b] | %+ | Test | Strains Positive | %+ |
|---|---|---|---|---|---|
| Catalase | 0/31 | 0 | Adonitol[f] | 0/1 | |
| Indole | 0/17 | 0 | Arabinose | 2/27 | 7 |
| Methyl Red | 13/13 | 100 | Cellobiose | 0/1 | |
| Voges-Proskauer | 0/1 | | Dextrin | 1/1 | |
| Nitrate Reduction | 2/31 | 6 | Dulcitol | 0/1 | |
| Nitrite Reduction | --------[c] | | Erythritol | 0/1 | |
| Esculin Hydrolysis | 13/14 | 93 | Fructose | 12/13 | 92 |
| Hippurate Hydrolysis | -------- | | Galactose | 1/1 | |
| Starch Hydrolysis | 17/17 | 100 | Glucose | 31/31 | 100 |
| Casein Hydrolysis | -------- | | Glycerol LA[g] | 1/13 | 7 |
| Milk-Acid, AR[d] | 4/18 | 22 | Thio | 8/14 | 57 |
| AC | 9/18 | 50 | Glycogen | 1/1 | |
| R | -------- | | Inositol | 1/1 | |
| NC | 3/14 | 21 | Inulin | 0/1 | |
| Gelatin Liquefaction | 0/13 | 0 | Lactose | 25/27 | 93 |
| Serum Liquefaction | -------- | | Maltose | 22/27 | 81 |
| DNase | -------- | | Mannitol | 3/31 | 10 |
| Acid Phosphatase | -------- | | Mannose | 9/17 | 53 |
| NH₄ from Arginine | -------- | | Melezitose | 0/1 | |
| Urease | 0/10 | 0 | Melibiose | 0/1 | |
| H₂S (B)HIA[e] | 6/23 | 26 | Raffinose | 0/17 | 0 |
| TSI | 10/10 | 100 | Rhamnose | 0/13 | 0 |
| | | | Ribose | 0/13 | 0 |
| Growth in | | | Salicin-LA | 0/12 | 0 |
| NaCL-4% | 0/12 | 0 | Thio | 5/14 | 36 |
| NaCL-6.5% | -------- | | Sorbitol | 0/1 | |
| | | | Starch | 26/27 | 96 |
| Bile-10% | -------- | | Sucrose | 22/27 | 81 |
| Bile 20% | 0/12 | 0 | Trehalose | 0/1 | |
| Bile 30% | -------- | | Xylose-LA | 2/17 | 12 |
| Bile 40% | -------- | | Thio | 8/14 | 57 |

[a] Data from Pine et al. (370); Georg et al. (143); Subgroup report (9).

[b] Number of strains positive/number of strains tested.

[c] -------- indicates no data was available for the test.

[d] Milk reactions include tests done in litmus milk and in nonhomogenized milk with iron filings. A=acid; AR=acid, reduced; AC=acid clot; R=reduced; NC=no change.

[e] (B)HIA=Brain Heart Infusion Agar or Heart Infusion Agar; TSI=Triple Sugar Iron Agar.

[f] This data combines the results from the use of different basal media with the media most frequently used being: 1) Liquid-Casitone (213), 2) Actinomyces Fermentation Base (BBL) and 3) Thioglycollate Fermentation Base (48).

[g] LA=results from the use of above medium 1 or 2; Thio=results from medium 3.

sorbitol. It is usually necessary to add 0.1-0.2% sterile serum to obtain adequate growth for the biochemical tests, especially fermentation tests. The only extensive published report on biochemical tests is that of Batty (17) which included 200 strains. She reported that a few strains fermented arabinose, galactose, mannitol and sucrose but not the other twelve sugars tested. These results are not included in the table as it is likely that many of the negative results were due to poor growth conditions.

**Table 6-3. Biochemical reactions of *Actinomyces odontolyticus*[a].**

| Test | Strains Positive[b] | %+ | Test | Strains Positive | %+ |
|---|---|---|---|---|---|
| Catalase | 0/44 | 0 | Adonitol[f] | 0/42 | 0 |
| Indole | 0/44 | 0 | Amygdalin | ---------[e] | |
| Methyl Red | 15/26 | 58 | Arabinose | 11/44 | 25 |
| Voges-Proskauer | 0/1 | | Cellobiose | 1/26 | 4 |
| Nitrate | 42/43 | 98 | Dextrin | 0/1 | |
| Nitrite | 0/26 | 0 | Dulcitol | 0/16 | 0 |
| Esculin Hydrolysis | 32/43 | 73 | Erythritol | 0/1 | |
| Starch Hydrolysis | 0/1 | | Fructose | 1/1 | |
| Hippurate Hydrolysis | 0/26 | 0 | Galactose | 0/1 | |
| | | | Glucose | 44/44 | 100 |
| Urease | 0/43 | 0 | Glycerol | 16/44 | 36 |
| | | | Glycogen | 15/26 | 58 |
| $H_2S$ – (B)HIA[c] | 5/42 | 12 | Inositol | 6/43 | 14 |
| TSI | 16/16 | 100 | Inulin | 1/42 | 2 |
| | | | Lactose | 12/26 | 46 |
| Gelatin Liquefaction | 0/42 | 0 | Maltose | 14/18 | 78 |
| Serum Liquefaction | ---------[e] | | Mannitol | 0/44 | 0 |
| Casein Hydrolysis | 7/26 | 27 | Mannose | 2/26 | 8 |
| Milk [-d] | | | Melezitose | 0/1 | |
|    A or AR | 8/16 | 50 | Melibiose | 0/1 | |
|    AC | 4/16 | 25 | Raffinose | 8/42 | 19 |
|    R | 0/16 | 0 | Rhamnose | 23/42 | 55 |
|    NC | 4/16 | 25 | Ribose | 21/26 | 81 |
| | | | Salicin | 21/43 | 49 |
| Ammonia from | | | Sorbitol | 0/42 | 0 |
|    Arginine | 0/26 | 0 | Starch | 41/44 | 93 |
| | | | Sucrose | 43/44 | 98 |
| | | | Trehalose | 25/42 | 60 |
| | | | Xylose | 31/44 | 70 |

[a] Data from Georg (9) and WVU unpublished data.

[b] Number of strains positive/number of strains tested.

[c] (B)HIA=Brain Heart Infusion Agar or Heart Infusion Agar; TSI=Triple Sugar Iron Agar.

[d] Milk reactions include tests done in litmus milk and in nonhomogenized milk with iron filings. A=acid; AR=acid, reduced; AC=acid clot; R=reduced; NC=no change.

[e] ------- indicates no data were available for the test.

[f] Thioglycollate Fermentation Base used for all fermentation tests.

*Actinomyces israelii* (Table 6-4). Because of its prominent involvement in human infections, this is the most widely studied of all the species of *Actinomyces*. In addition to the references used for the table, others include: Erikson (115), Holm (202, 203), Lentze (277), Negroni and Bonfiglioli (338), Ludwig and Sullivan (292), King and Meyer (243) and Georg (136).

A. *israelii* hydrolyzes esculin and is urease negative; most strains are methyl-red positive. About half reduce nitrate or hydrolyze starch with a small zone of clearing in 7-14 days. Acid phosphatase production can be tested using disodium-p-nitrophenyl phosphate at pH 4.5 and incubating duplicate tests at 37 and 50 C (210). Eighty percent of strains tested were positive at 50 C, but some strains of A. *naeslundii* and A. *viscosus* were also positive. Even though reproducibility is poor, the test deserves further use and study. A method for determining ammonia production from arginine has been developed based on procedures described by Thornley (457)

**Table 6-4.** Biochemical reactions of *Actinomyces israelii*[a].

| Test | Strains Positive[b] | %+ | Test | Strains Positive | %+ |
|---|---|---|---|---|---|
| Catalase | 0/266 | 0 | Adonitol[e] | 0/16 | 0 |
| Indole | 0/251 | 0 | Amygdalin | --------- | |
| Methyl Red | 70/80 | 88 | Arabinose[g] | | |
| Voges-Proskauer | 6/196 | 3 | type 1[f] | 66/83 | 80 |
| Nitrate | 143/263 | 54 | type 2 | 0/25 | 0 |
| Nitrite | 0/21 | 0 | Cellobiose[g] | 69/81 | 85 |
| | | | Dextrin | 1/1 | |
| Esculin Hydrolysis | 47/49 | 96 | Dulcitol[h] | 0/1 | |
| Starch Hydrolysis | 26/63 | 41 | Erythritol | 0/1 | |
| Hippurate Hydrolysis | 1/15 | 6 | Fructose | 1/1 | |
| | | | Galactose | 0/1 | |
| Urease | 0/49 | 0 | Glucose | 266/266 | 100 |
| | | | Glycerol[g] | 0/108 | 0 |
| H$_2$S – (B)HIA[c] | 0/27 | 0 | Glycogen | 0/81 | 0 |
| TSI | 27/27 | 100 | Inositol[g] | 178/264 | 67 |
| | | | Inulin | 61/174 | 35 |
| Gelatin Liquefaction | 0/229 | 0 | Lactose[g] | 237/266 | 89 |
| Serum Liquefaction | 0/16 | 0 | Maltose | 183/185 | 98 |
| Casein Hydrolysis | 4/16 | 25 | Mannitol[g] | 184/266 | 69 |
| Milk-A-AR[d] | 64/179 | 36 | Mannose | 50/64 | 78 |
| AC | 62/179 | 35 | Melezitose | 15/70 | 21 |
| R | 36/179 | 20 | Melibiose | 1/1 | |
| NC | 17/179 | 9 | Raffinose[g] | 132/239 | 55 |
| | | | Rhamnose | 7/78 | 9 |
| Growth in NaCl | | | Ribose | 138/175 | 79 |
| 4.0% | 4/100 | 4 | Salicin[g] | 176/266 | 66 |
| 6.5% | 0/100 | 0 | Sorbitol | 11/71 | 15 |
| | | | Starch | 43/108 | 40 |
| Growth in 10% bile | 0/95 | 0 | Sucrose | 200/202 | 99 |
| | | | Trehalose | 17/17 | 100 |
| Acid Phosphatase | 12/15 | 80 | Xylose | 250/266 | 94 |
| Ammonia from Arginine | 13/13 | 100 | | | |
| DNase | 0/16 | 0 | | | |

[a] Data from Howell et al. (217); Slack et al. (424); Brock and Georg (48); WVU unpublished data; Subgroup study (9).

[b] Number of strains positive/number of strains tested.

[c] (B)HIA=Brain Heart Infusion Agar or Heart Infusion Agar; TSI=Triple Sugar Iron Agar.

[d] Milk reactions include tests done in litmus milk and in nonhomogenized milk with iron filings. A=acid; AR=acid, reduced; AC=acid clot; R=reduced; NC=no change.

[e] This data combines the results from the use of different basal media with the media most frequently used being: 1) Liquid-Casitone (213), 2) Actinomyces Fermentation Base (BBL) and 3) Thioglycollate Fermentation Base (48).

[f] Type 1 = serotype 1; type 2 = serotype 2.

[g] See Table 6-5.

[h] According to Georg (136), most strains ferment dulcitol.

and Niven et al. (341). All thirteen strains of *A. israelii* tested produced ammonia in 7-14 days. Indications are that this test may be specific for *A. israelii* as, of the other species tested, only an occasional strain of *A. naeslundii* gave a weak positive test. However, the test needs refinement to improve growth conditions and shorten the incubation time.

*A. israelii* ferments glucose, maltose, sucrose, trehalose and xylose, while adonitol, glycerol and glycogen are uniformly negative. With some carbohydrates, the results in Table 6-4 may be misleading as to the amount of strain variation which occurs. Table 6-5 shows the results of five separate studies for several carbohydrates. Howell et al. (213) found that fewer strains fermented arabinose, raffinose and salicin than did the other workers. In our studies, about 50% of the strains have fermented mannitol regardless of the basal medium, but other studies find a higher percentage (75 to 100%) of strains producing acid from mannitol. We also find fewer strains fermenting inositol when our results on dental isolates are compared with those of Brock and Georg (48). As yet, we cannot explain these discrepancies, but they are not due to differences in the basal medium. The results of fermentation tests done under strictly anaerobic conditions (201) agree in general with those obtained in thioglycollate fermentation base except that arabinose was usually negative.

The question of the fermentation of arabinose deserves further consideration. Holm (203) divided strains of *Actinomyces* into two groups on the basis of arabinose fermentation and $CO_2$ requirements. Subsequent studies (9, 48, 424) have shown that his arabinose-negative group II strains could be classified as *A. israelii*, serotype 2. The correlation between arabinose fermentation and serotype is not absolute as some strains of serotype 1 are arabinose negative (424; dental isolates Table 6-5). Comparison of the studies shown in Table 6-5 indicates that arabinose fermentation is influenced by the basal medium since Howell et al. (213) found that only 10% of 158 strains produced acid in the Liquid Casitone-Starch Medium. Our studies with the suggested neotype strain, ATCC 12102, can be used as a specific case of the influence of basal

**Table 6-5. Fermentation of selected carbohydrates by *A. israelii*
to show the influence of media.**

| Carbo-hydrates | Howell[a] et al. (213)  % | Slack[b] et al. (424)  % | Brock and Georg[c] (48)  % | WVU[d] Dental Isolates  % | VPI[e] (201) |
|---|---|---|---|---|---|
| Arabinose | 16/158[f]  10 | 41/64  64 | 13/27  48 | 12/17  71 | -[w] |
| Serotype 1 | ----------[g] | 41/53  77 | 13/13  100 | 12/17  71 | ----- |
| Serotype 2 | ---------- | 0/11  0 | 0/14  0 | 0/0  0 | ----- |
| | | | | | |
| Cellobiose | ---------- | 54/64  84 | -------- | 15/17  88 | A |
| Glycerol | 0/158  0 | 0/64  0 | 0/27  0 | 0/17  0 | - |
| Inositol | 99/158  63 | 39/62  63 | 27/27  100 | 13/17  76 | A |
| Lactose | 139/158  88 | 57/64  89 | 26/27  96 | 15/17  88 | A |
| Mannitol | 118/158  75 | 31/64  48 | 27/27  100 | 8/17  47 | A |
| Raffinose | 56/158  35 | 62/64  97 | ---------- | 14/17  82 | A[w] |
| Salicin | 70/158  44 | 63/64  98 | 27/27  100 | 16/17  94 | A |

[a] Basal medium—Liquid Casitone Starch with pyrogallol seal.

[b] Basal medium—Actinomyces Fermentation Broth (BBL) with pyrogallol seal.

[c] Basal medium—Thioglycollate Fermentation Base.

[d] Medium is the same as c. These are results from our own laboratory with fresh isolates from human dental plaque or calculus.

[e] Prereduced anaerobic media. The number of strains was not given. - = negative; A = 90% or more acid;-[w] or A[w] = occasional reaction.

[f] Number of positive/number of strains tested.

[g] ---- indicates no data available for the test.

medium. This strain consistently ferments arabinose when TFB is used and is consistently negative when AFB is used. This also illustrates that medium effects may be strain variable since 77% of the serotype 1 strains we studied did ferment arabinose in AFB. It should also be pointed out that with other sugars the results with these two media were comparable when used with *A. israelii*. Anaerobiosis may also be a factor in arabinose fermentation since VPI reports that most strains of *A. israelii* do not ferment arabinose.

*Actinomyces naeslundii* (Table 6-6). Over 90% of the strains hydrolyze esculin and reduce nitrate to nitrite. A few strains reduce nitrite, but reproducibility of this test is poor. The hydrolysis of starch and casein is variable as is the production of acid phosphatase. Neither lecithinase nor lipase is produced (201). The results of tests done under strictly anaerobic conditions (201) are in general agreement with those listed in the table as are the results of Holmberg and Hallender (207). In addition, this latter study reports that *A. naeslundii* does not hydrolyze casein, tyrosine or xanthine and does not grow on Rogosa's medium or with an inorganic source of nitrogen. The only major discrepancy between this study and the results in the table was in the urease test. All strains of *A. naeslundii* were urease positive when tested on Christensen's urea agar. A recent report by Scharfen (407) has confirmed the results of Holmberg and Hallender and demonstrated that differences in urease test results are medium dependent. He found that *A. naeslundii* was urease positive on urea agar (Christensen's) and in tryptone soya broth with urea but was urease negative in thioglycollate urea broth.

*A. naeslundii* ferments glucose, glycerol, maltose, mannose, melibiose, raffinose and sucrose, and most strains ferment lactose, salicin and trehalose when tests are done in TFB. Adonitol, arabinose, mannitol, melezitose and xylose are not fermented. Table 6-7 shows the variations recorded by different authors for certain sugar fermentations. The fermentation of glycerol is highly medium dependent. Almost 90% of the strains will ferment glycerol when TFB is used, but they are negative with AFB (146). *A. naeslundii* also ferments glycerol in 1% peptone agar deeps (207) but is negative in casitone-starch broth (213) and in peptone-yeast extract broth under strictly anaerobic conditions (201). It is difficult to determine whether nutritional factors, degree of anaerobiosis or both influence glycerol fermentation by *A. naeslundii*. It should be remembered that *A. israelii* is negative in glycerol under all these various conditions. Variability with inositol, inulin, ribose and starch may be related to methodology as well as to strain variation. One as yet unexplained difference occurred using the same test conditions as Coleman et al. (78). They reported that 75% of the strains fermented starch while we found that only 29% of the strains produced acid.

*Actinomyces viscosus* (Table 6-8). This is the only *Actinomyces* species which is catalase positive. It should be noted that anaerobically grown cultures may be catalase negative if tested immediately upon removal from the jar. Thus, such cultures should be exposed to the air for 30 minutes before being tested for catalase. Bier and Araujo (27) reported two catalase negative strains as *A. viscosus*. These strains have been included in the table because they could not be eliminated without dropping the entire fifty strains. If these strains are catalase negative under aerobic conditions, then under present criteria they probably should be classified as *A. naeslundii*.

About 95% of strains of *A. viscosus* reduce nitrate to nitrite when Indol-Nitrate medium (BBL) is used (27, 142, 146). The table includes the results of Howell and Jordan (211) who found that only ten of twenty-three reduced nitrate when tested in enriched thioglycollate broth and that all strains were negative in a modified liquid-casitone medium. It is interesting to note that strain ATCC 19246 (WVU 371), which we have used as a prototype human isolate, is nitrate negative. In trypticase soy broth with 0.01% $KNO_2$, about 50% of our strains and over 90% of strains tested by Bier and Araujo were nitrite positive. Thus, this test is not reliable for separating *A. viscosus* from *R. dentocariosa* as suggested by Brown et al. (52). In contrast to our results and those of VPI, Bier and Araujo (27) reported that all fifty of their strains hydrolyzed hippurate. The reason for this difference is not apparent but probably involves methodology. As with *A. naeslundii,* urease reactions are medium dependent. All strains are urease negative when tested in thioglycollate urea broth (Table 6-8) but positive on urea agar (Christensen's) and tryptone soya broth with urea (207, 407).

**Table 6-6. Biochemical reactions of *Actinomyces naeslundii*[a].**

| Test | Positive[b] Reaction | %+ | Test | Positive Reaction | %+ |
|---|---|---|---|---|---|
| Catalase | 0/108 | 0 | Adonitol[e] | 0/26 | 0 |
| Indole | 0/104 | 0 | Amygdalin | -------- | |
| Methyl Red | 50/55 | 91 | Arabinose | 2/108 | 2 |
| Voges-Proskauer | 0/50 | 0 | Cellobiose | 31/55 | 56 |
| Nitrate | 99/108 | 92 | Dextrin | 1/1 | |
| Nitrite | 9/50 | 18 | Dulcitol | 0/12 | 0 |
| | | | Erythritol | 0/1 | |
| Esculin Hydrolysis | 62/67 | 93 | Fructose | 1/1 | |
| Starch Hydrolysis | 14/18 | 77 | Galactose | 1/1 | |
| Hippurate Hydrolysis | 0/13 | 0 | Glucose | 108/108 | 100 |
| | | | Glycerol | 60/108 | 55 |
| Urease | 0/67 | 0 | LA[f] | 0/41 | 0 |
| | | | Thio | 60/67 | 90 |
| H$_2$S – (B)HIA[c] | 9/25 | 36 | Glycogen | 7/55 | 13 |
| TSI | 5/5 | 100 | Inositol | 94/108 | 87 |
| | | | Inulin | 35/71 | 49 |
| Gelatin Liquefaction | 0/105 | 0 | Lactose | 89/108 | 82 |
| Serum Liquefaction | 0/13 | 0 | Maltose | 49/53 | 92 |
| Casein Hydrolysis | 5/13 | 38 | Mannitol | 0/108 | 0 |
| Milk A,AR[d] | 23/52 | 44 | Mannose | 13/13 | 100 |
| AC | 23/52 | 44 | Melezitose | 0/31 | 0 |
| R | 0/52 | 0 | Melibiose | 5/5 | 100 |
| NC | 6/52 | 12 | Raffinose | 102/108 | 94 |
| | | | Rhamnose | 11/62 | 17 |
| DNase | 0/13 | 0 | Ribose | 49/91 | 54 |
| Acid Phosphatase | 4/13 | 31 | Salicin | 53/67 | 79 |
| NH$_4$ from Arginine | 0/13 | 0 | Sorbitol | 3/13 | 23 |
| | | | Starch | 25/67 | 37 |
| Growth in | | | Sucrose | 108/108 | 100 |
| 4% NaCl | 8/28 | 29 | Trehalose | 57/67 | 85 |
| 6.5% NaCl | 0/28 | 0 | Xylose | 2/108 | 2 |
| 10% bile | 17/28 | 61 | | | |
| 20% bile | 15/28 | 54 | | | |
| 30% bile | 11/28 | 39 | | | |
| 40% bile | 3/27 | 11 | | | |

[a] Data from: Howell et al. (213); Coleman et al. (78); Gerencser and Slack (146); Subgroup study (9); WVU unpublished data.

[b] Number of strains positive/number of strains tested.

[c] (B)HIA=Brain Heart Infusion Agar or Heart Infusion Agar; TSI=Triple Sugar Iron Agar.

[d] Milk reactions include tests done in litmus milk and in nonhomogenized milk with iron filings. A=acid; AR=acid, reduced; AC= acid clot; R=reduced; NC=no change.

[e] This data combines the results from the use of different basal media with the media most frequently used being: 1) Liquid-Casitone (213), 2) Actinomyces Fermentation Base (BBL) and 3) Thioglycollate Fermentation Base (146).

[f] LA=results from the use of above medium 1 or 2; Thio=results from medium 3.

Carbohydrate fermentation results vary and are probably medium dependent. Thioglycollate Fermentation Base appears to give the most consistent results of media we have used (146). All strains ferment glucose, maltose and sucrose; most ferment glycerol; and none ferment arabinose,

Table 6-7. Variation in fermentation of certain sugars by *Actinomyces naeslundii*.

| Source | Glycerol +/Tested[a] | % | Inositol +/Tested | % | Inulin +/Tested | % | Ribose +/Tested | % | Starch +/Tested | % |
|---|---|---|---|---|---|---|---|---|---|---|
| Howell et al. (213) | 0/41 | 0 | 31/41 | 76 | 13/41 | 32 | 8/41 | 20 | ---d | |
| Coleman et al. (78 ) | 11/12 | 92 | 11/12 | 92 | 10/12 | 83 | ------ | | 9/12 | 75 |
| WVU unpublished | 49/55 | 89 | 52/55 | 95 | 12/18 | 67 | 41/50 | 82 | 16/55 | 29 |
| VPI (201)[b] | _w | | v | | a⁻ | | _w | | v | |
| Holmberg and Hallender (207)[c] | + | | + | | --- | | --- | | --- | |

[a]Number of strains positive/number of strains tested.

[b] a =90% or more +; − = 90% or more negative: v = + or −; −w = occasional reaction.

[c]+ = 90% or more +; = 90% or more negative

d — indicates no data were available for the test.

adonitol or xylose. In a solid medium with an infusion base and Andrades indicator, the results were similar except that arabinose (48/50) was routinely fermented (27). If medium is responsible for this difference, only the one sugar was affected. Despite the preferential aerobic growth of *A. viscosus,* fermentation results were similar when done under strictly anaerobic conditions (201) except that glycerol was negative. In the study of Howell and Jordan (211), none of these sugars, not even glucose, was fermented by all strains in the casitone-starch medium. This may well have been due to poor growth as we obtained similar results in AFB, which is a modification of the casitone-starch medium.

*Actinomyces humiferus* (Table 6-9). The biochemical reactions for this species have only been reported by Gledhill and Casida (160). *A. humiferus* resembles the other *Actinomyces* in being catalase and indole negative. It is nitrate and urease negative and about one-half the strains hydrolyze gelatin on gelatin agar plates. Starch is hydrolyzed with wide zones of hydrolysis, and most strains also hydrolyze casein and esculin. A large number of carbohydrates (as listed in the table) are fermented with the production of acid but no gas.

*Actinomyces suis* (Table 6-10). Swine isolates were first reported and named by Grässer (173, 174). Additional swine isolates were described by Biever (28) and more recently by Franke (128). The results from these three reports with a total of eighteen strains are listed in Table 6-10. These are listed separately as it is difficult to determine whether or not these are actually the same organism. It is evident that there is need for someone to collect and isolate a greater number of strains and complete a thorough study of the morphological, biochemical and serological characteristics of this organism. Morphologically, *A. suis* resembles the other species. It is catalase and indole negative and ferments a wide variety of carbohydrates.

This discussion of the biochemical reactions of *Actinomyces* brings out several factors which must be considered in studying these organisms. Techniques for doing biochemical tests have improved but are obviously not optimal and reproducibility remains a problem in some cases.

**Table 6-8. Biochemical reactions of *Actinomyces viscosus*[a].**

| Test | Positive[b] Reaction | %+ | Test | Positive Reaction | %+ |
|---|---|---|---|---|---|
| Catalase | 139/141 | 99 | Adonitol[f] | 0/34 | 0 |
| Indole | 0/133 | 0 | Amygdalin | -------- | – |
| Methyl Red | 97/103 | 94 | Arabinose | 0/91 | 0 |
| Voges-Proskauer | 0/100 | 0 | | 48/50[e] | 96 |
| Nitrate | 114/131 | 87 | Cellobiose | 8/53 | 15 |
| Nitrite | 58/102 | 57 | Dextrin | 0/2 | |
| Esculin | 101/106 | 95 | Dulcitol | 0/7 | 0 |
| Starch Hydrolysis | 80/93 | 86 | Erythritol | 0/2 | |
| Hippurate Hydrolysis | 1/17 | 6 | Fructose | 30/35 | 86 |
| | 50/50[e] | 100 | Galactose | 24/35 | 69 |
| Urease | 0/56 | 0 | Glucose | 135/141 | 96 |
| $H_2S$ — (B)HIA[c] | 0/99 | 0 | Glycerol | 97/136 | 71 |
| TSI | 26/26 | 100 | LA[g] | 0/35 | 0 |
| Gelatin Liquefaction | 0/132 | 0 | Thio | 97/101 | 96 |
| Serum Liquefaction | 0/17 | 0 | Glycogen | 9/53 | 17 |
| Casein Hydrolysis | 7/39 | 18 | Inositol | 123/136 | 90 |
| Milk[d] | | | Inulin | 74/103 | 72 |
| A,AR | 18/28 | 64 | Lactose | 33/141 | 23 |
| AC | 10/28 | 36 | Maltose | 88/90 | 98 |
| R | 0/28 | 0 | Mannitol | 3/141 | 2 |
| NC | 0/28 | 0 | Mannose | 33/37 | 89 |
| DNase | 3/17 | 18 | Melezitose | 0/2 | |
| Acid Phosphatase | 9/17 | 53 | Melibiose | 18/24 | 75 |
| $NH_4$ from Arginine | 0/17 | 0 | Raffinose | 133/136 | 98 |
| | | | Rhamnose | 2/64 | 3 |
| Growth in 2% NaCl | 40/50 | 80 | Ribose | 23/64 | 36 |
| 4% NaCl | 34/81 | 42 | Salicin | 108/141 | 77 |
| 10% bile | 22/81 | 27 | Sorbitol | 0/37 | 0 |
| 20% bile | 17/31 | 55 | Starch | 34/53 | 64 |
| 30% bile | 8/31 | 26 | Sucrose | 140/141 | 99 |
| 40% bile | 5/31 | 16 | Trehalose | 89/103 | 86 |
| | | | Xylose | 0/141 | 0 |
| Tyrosine Hydrolysis | 0/24 | 0 | | | |
| Xanthine Hydrolysis | 0/24 | 0 | | | |

[a] Data from: Howell and Jordan (211); Gerencser and Slack (146); Bier and Araujo (27); Subgroup report (9); WVU unpublished data.

[b] Number of strains positive/number of strains tested.

[c] (B)HIA=Brain Heart Infusion Agar or Heart Infusion Agar; TSI=Triple Sugar Iron Agar.

[d] Milk reactions include tests done in litmus milk and in nonhomogenized milk with iron filings. A=acid; AR=acid, reduced; AC=acid clot; R=reduced; NC=no change.

[e] Data from Bier and Araujo (27).

[f] This data combines the results from the use of different basal media with the media most frequently used being: 1) Liquid-Casitone (213), 2) Actinomyces Fermentation Base (BBL) and 3) Thioglycollate Fermentation Base (146).

[g] LA = results from the use of above medium 1 or 2; Thio = results from medium 3.

**Table 6-9. Biochemical reactions of *Actinomyces humiferus*[a].**

| Test | Strains Positive[b] | %+ | Test | Strains Positive | %+ |
|---|---|---|---|---|---|
| Catalase | 0/27 | 0 | Adonitol[e] | 0/27 | 0 |
| Indole | 0/27 | 0 | Arabinose | 26/27 | 96 |
| Methyl Red | 27/27 | 100 | Cellobiose | 18/27 | 67 |
| Voges-Proskauer | 4/27 | 15 | Dextrin | 27/27 | 100 |
| Nitrate Reduction | 0/27 | 0 | Dulcitol | 0/27 | 0 |
| Nitrite Reduction | ------[d] | | Erythritol | ------ | |
| Esculin Hydrolysis | 24/27 | 89 | Fructose | 27/27 | 100 |
| Hippurate Hydrolysis | ------ | | Galactose | 25/27 | 93 |
| Starch Hydrolysis | 27/27 | 100 | Glucose | 27/27 | 100 |
| Casein Hydrolysis | 23/27 | 85 | Glycerol | 20/27 | 74 |
| Milk-Acid, AR[c] | 22/27 | 81 | Glycogen | ------ | |
| AC | 0/27 | 0 | Inositol | 2/27 | 7 |
| R | 0/27 | 0 | Inulin | 1/27 | 4 |
| NC | 5/27 | 18 | Lactose | 8/27 | 29 |
| Gelatin Liquefaction[f] | 0/27 | 0 | Maltose | 26/27 | 96 |
| Serum Liquefaction | ------ | | Mannitol | 25/27 | 93 |
| DNase | ------ | | Mannose | 26/27 | 96 |
| Acid Phosphatase | ------ | | Melezitose | 26/27 | 96 |
| $NH_4$ from Arginine | 0/27 | 0 | Melibiose | 25/27 | 93 |
| Urease | 0/27 | 0 | Raffinose | ------ | |
| $H_2S$, BHIA | 18/27 | 67 | Rhamnose | 26/27 | 96 |
| Growth in | | | Ribose | 3/27 | 11 |
| NaCl 4% | 14/27 | 52 | Salicin | 21/27 | 78 |
| Bile 1% | 9/27 | 33 | Sorbitol | 1/27 | 4 |
| 3% | 0/27 | 0 | Starch | 27/27 | 100 |
| | | | Sucrose | 27/27 | 100 |
| | | | Trehalose | 19/27 | 70 |
| | | | Xylose | 24/27 | 89 |

[a] Data from Gledhill and Casida (160).

[b] Number of strains positive/number of strains tested.

[c] Milk reactions include tests done in litmus milk and in nonhomogenized milk with iron filings. A=acid; AR=acid, reduced; AC=acid clot; R=reduced; NC=no change.

[d] ----- indicates no data available for the test.

[e] Basal medium was tryptone-casamino acids broth.

[f] 12/27 (44%) hydrolyze gelatin on gelatin agar plates.

Despite these problems and the wide range of strain variation, biochemical tests can be used for identifying *Actinomyces* if the pattern of reactions is considered rather than single tests.

## E. SEROLOGY

**Introduction.** Early serological investigations on *Actinomyces* and other actinomycetes using the agglutination test have been reviewed by Slack et al. (426). Although used extensively, agglutination tests were never truly satisfactory, especially for *A. israelii,* because of the difficulty in producing a homogeneous antigen which would not autoagglutinate. Even with these difficulties, Erikson (115) showed a clear-cut serological difference between *A. israelii* and *A. bovis.* Holm (202) and Lentze (277) demonstrated two serological groups among human *Actinomyces*

Table 6-10. Biochemical reactions of *Actinomyces suis*.

| Test | Grässer[a] No+ | %+ | Biever[b] No+ | %+ | Franke[c] No+ | %+ |
|---|---|---|---|---|---|---|
| No. strains | 9 | | 7 | | 8 | |
| Catalase | ------[d] | | 0 | 0 | 0 | 0 |
| Indole | ------ | | 0 | 0 | 0 | 0 |
| Methyl Red | ------ | | 2 | 29 | 8 | 100 |
| Voges-Proskauer | ------ | | 0 | 0 | 0 | 0 |
| Nitrate Reduction | ------ | | 1 | 14 | 8 | 100 |
| Nitrite Reduction | ------ | | ------ | | ------ | |
| Esculin Hydrolysis | ------ | | ------ | | ------ | |
| Hippurate Hydrolysis | ------ | | ------ | | ------ | |
| Casein Hydrolysis | ------ | | ------ | | ------ | |
| Milk-Acid, AR[f] | ------ | | ------ | | ------ | |
| AC | | | | | | |
| R | | | | | | |
| NC | | | | | | |
| Gelatin Liquefaction | ------ | | 0 | 0 | 0 | 0 |
| DNase | ------ | | ------ | | ------ | |
| Acid Phosphatase | ------ | | ------ | | ------ | |
| NH4 from Arginine | ------ | | ------ | | ------ | |
| Urease | ------ | | ------ | | ------ | |
| H2S — (B)HIA | ------ | | 1 | 14 | 0 | 0 |
| Adonitol | 7 | 78 | 7 | 100 | v[e] | – |
| Arabinose | 8 | 89 | ------ | | 0 | 0 |
| Cellobiose | ------ | | ------ | | ------ | |
| Dulcitol | 0 | 0 | ------ | | 0 | 0 |
| Esculin | 5/5 | 100 | ------ | | ------ | |
| Fructose | 9 | 100 | ------ | | 8 | 100 |
| Galactose | 9 | 100 | ------ | | 8 | 100 |
| Glucose | 9 | 100 | 7 | 100 | 8 | 100 |
| Glycerol | 6 | 67 | 7 | 100 | 7 | 88 |
| Inositol | ------ | | ------ | | 8 | 100 |
| Inulin | 9 | 100 | ------ | | 8 | 100 |
| Lactose | 9 | 100 | ------ | | 8 | 100 |
| Maltose | 9 | 100 | ------ | | 8 | 100 |
| Mannitol | 5 | 56 | 6 | 86 | v | |
| Mannose | ------ | | ------ | | 8 | 100 |
| Raffinose | 9 | 100 | 6 | 86 | 8 | 100 |
| Rhamnose | 4 | 44 | ------ | | 0 | 0 |
| Ribose | ------ | | ------ | | 7 | 88 |
| Salicin | 9 | 100 | ------ | | 8 | 100 |
| Sorbitol | 1 | 11 | ------ | | 0 | 0 |
| Starch | 9 | 100 | 6 | 86 | 8 | 100 |
| Sucrose | 8 | 89 | ------ | | 8 | 100 |
| Trehalose | 9 | 100 | ------ | | 8 | 100 |
| Xylose | 9 | 100 | 6 | 86 | 7 | 88 |

[a]Grässer (174) 9 S and R colony types from three strains. Fermentation Base – Peptone water + 10% horse serum + Bromthymol blue.

[b]Biever, M.S. Thesis, South Dakota State University, 1967. Fermentation Base – Thioglycollate without glucose or indicator + yeast extract + Bromthymol blue.

[c]Franke (128). Fermentation Base – Thioglycollate + Serum Medium + Phenol Red Indicator.

[d]No data available.

[e]Variable reactions, numbers not given.

[f]Milk reactions include tests done in litmus milk and in nonhomogenized milk with iron filings. A=acid; AR=acid, reduced; AC=acid clot; R=reduced; NC=no change.

isolates. As cultures of these second serological groups are not available, it is impossible to determine their identity. However, from the biochemical reactions listed by Lentze, it is most likely that his were anaerobic corynebacteria.

Immunodiffusion (ID), fluorescent antibody (FA) and cell-wall agglutination tests (CWA) have now been used successfully in studying *Actinomyces* in a number of laboratories. This discussion will be concerned first with the use of serological tests for identification and classification followed by a brief review of reports on the chemical composition of the antigens of *Actinomyces*.

**Serological Identification of *Actinomyces*.** Serological differences between *Actinomyces* species which correlate with other speciation methods can be demonstrated with ID, FA and CWA procedures. Of these, immunofluorescent techniques have been used most widely. The FA technique was first applied to the serological grouping of *Actinomyces* by Slack et al. (427) and was used later to revise the grouping (420) to eliminate organisms then recognized as anaerobic diphtheroids. This serological grouping (420, 421) is shown in Table 6-11. The serological groups, designated by capital letters, so far correspond to the *species* so that species names rather

Table 6-11. Serological groups and types of *Actinomyces*.

| Species | Serological Group | Serotypes |
|---|---|---|
| *A. bovis* | A | 1, 2 |
| *A. naeslundii* | B | 1, 2 |
| *A. eriksonii*[a] | C | 1, 2 |
| *A. israelii* | D | 1, 2 |
| *A. odontolyticus* | E | 1, 2 |
| *A. viscosus* | F | 1, 2 |

[a] To be reclassified as *Bifidobacterium*.

than serogroup designations are usually used. Two serotypes have been established for each species, but evidence suggests that additional serotypes exist. The revised serological grouping has been confirmed in subsequent studies; however, *A. eriksonii* should be eliminated since it will be reclassified as *Bifidobacterium*. The serological cross-reactions between groups and types are shown in Table 6-12. The table has been arranged to emphasize serological relationships rather than listing the groups alphabetically, and *B. eriksonii* has been omitted.

**Preparation of antisera.** Numerous methods for the preparation of antigens and immunization schedules have been used. Currently, in this laboratory, the antigens are live, washed, whole-cell suspensions. Rabbits are injected intravenously with 1.0 ml of antigen on each Monday, Wednesday and Friday for 4 weeks and trial bled 7 days after the final injection. This will usually provide FA titers of 1:20 or better, and such rabbits are given one week of booster injections and then bled out. Additional immunization does not seem to improve the titer, and occasionally a rabbit will not respond and is discarded.

For conjugation, the antiserum is precipitated with 35% $NH_4SO_4$ according to the procedure of Hebert et al. (194) and conjugated with fluorescein isothiocyanate (FITC) at a ratio of 0.05 mg FITC/mg of protein. The conjugates are titered using two-fold dilutions in FTA buffer (BBL) to stain fixed smears of the immunizing antigen. The highest dilution giving 4 + fluorescence is designated as the titer. For routine use this conjugate is then diluted to one two-fold dilution less than the determined titer—this is designated as the "working" or "diagnostic" titer. For example, the working titer of a conjugate with a titer of 1:80 would be 1:40. When necessary, conjugates are sorbed with cross-reacting antigens by incubating 1.0 ml of the conjugate with 0.1 ml of living packed cells of the sorbing strain for 1 hr. at 55 C with intermittent mixing. The mixture is refrigerated overnight and centrifuged to remove the cells. Some sera will have to be sorbed two to three times with fresh living cells to remove all cross-reactions.

Recently, Holmberg and Forsum (206) have compared IgG conjugates with various F/P ratios. They reported that IgG fractions containing 10 mg/ml of protein and conjugated at a

|  |  |  | *A. bovis* B | | *A. odontolyticus* E | | *A. israelii* D | | *A. naeslundii* [b] A | | *A. viscosus* F | |
|---|---|---|---|---|---|---|---|---|---|---|---|---|
| **Group** | **Species** |  | 1 | 2 | 1 | 2 | 1 | 2 | 1 | 2 | 1 | 2 |
| B | *A. bovis* | 1 | + | − | − | − | − | − | − |  | − | − |
|  |  |  | − | + | − | − | − | − | − |  | − | − |
| E | *A. odontolyticus* | 1 | − | − | + | (+) | − | − | − |  | − | − |
|  |  | 2 | − | − | (+) | + | − | − | − |  | (+)or + | − |
| D | *A. israelii* | 1 | − | − | − | − | + | (+) | (+) |  | (+) | (+) |
|  |  | 2 | − | − | − | − | (+) | + | − |  | − | − |
| A | *A. naeslundii* | 1[b] | − | − | − | − | (+) | − | + |  | (+) | (+) |
| F | *A. viscosus* | 1 | − | − | − | (+) | − | − | − |  | + | − |
|  |  | 2 | − | − | − | − | (+) | − | (+) |  | (+) | + |

*Antisera*
Species, Group and Serotype

*Antigens*

[a] + = 3–4 + fluorescence at the diagnostic titer of the antiserum
(+) = 3–4 + fluorescence at titers lower than the diagnostic titer of the antiserum − these cross-reactions are strain variable
− = negative fluorescence
[b] a second serotype of *A. naeslundii* has been reported by Bragg et al. (46), but its relationship to other species has not been completely studied.

dye to protein ratio of 10 or less gave high specific staining titers with little or no nonspecific staining.

**Antigens for FA.** These antigens consist of fixed smears of whole cells. There is apparently little difference in the staining properties of cells grown on different complex media, but this point has not been systematically investigated. Young cells generally stain better than old cells, but killed cells stain as well as living ones.

**Results with FA.** The first part of this discussion refers primarily to the results shown in Table 6-12. Most of the interspecies cross-reactions can be eliminated by dilution of the antiserum, although sorption may occasionally be required. The table is based on our studies, but these results have been confirmed in other laboratories (136, 206, 264). Different workers report varying degrees of cross-reaction between species and serotypes within species, but the correlation between species identification and serological grouping has always been found.

*A. bovis,* group B, shows no cross-reactions with other species or between serotypes, but minor antigenic differences may exist among the strains within a serotype. Among strains studied to date, serotype 1 strains produce smooth nonfilamentous microcolonies, while serotype 2 strains produce filamentous microcolonies.

*A. odontolyticus,* group E, has been studied less intensively than the other species, and the second serotype reported by Slack and Gerencser (421) has not yet been confirmed. Strain variable cross-reactions occur between the serotypes, and serotype 2 cross-reacts with *A. viscosus,* 1. Holmberg and Forsum (206) found no cross-reactions between *A. odontolyticus* and other species, and *A. odontolyticus* in plaque did not stain with *A. viscosus,* 1 antiserum.

The other three species, *A. israelii*, *A. naeslundii* and *A. viscosus*, each contain two serotypes and show serological relationships with each other. Cross-reactions, both between serotypes in a species and between species, are generally low titered and show varying degrees of strain variation.

Two serotypes of *A. israelii*, group D, were first reported by Lambert et al. (263) and then confirmed by others (33, 49, 424). Brock and Georg (49) found only a one-way cross-reaction between the serotypes in which type 1 antiserum stained type 2 cultures, while Slack et al. (424) and Holmberg and Forsum (206) found reciprocal cross-reactions. *A. israelii*, 1 strains exhibit antigen variation within the serotype. For example, *A. israelii* (ATCC 10048) is not identical antigenically with *A. israelii* (ATCC 12102) (42, 49, 424).

**Interspecies Cross-Reactions Demonstrable with FA (Figure 6-79).** Interspecies cross-reactions do occur, but they are usually low-grade and strain variable. As already explained above, species-specific antiserum can be prepared by sorption, although simple dilution is usually sufficient to eliminate these cross-reactions. These are described in Figure 6-79 on the basis of the reactions with the unsorbed antiserum of *A. israelii*, 1 and 2, *A. naeslundii*, 1 and *A. viscosus*, 1 and 2. A diagram is given for each, followed by a short discussion. Dotted lines are used to show the inter-relationships to emphasize that the reactions are usually low-grade.

**Immunodiffusion (ID) Reactions.** ID tests were first used by King and Meyer (244) to show antigenic relationships between *Actinomyces* species and to show that they did not cross-react with anaerobic diphtheroids. They used antisera against whole cells and soluble antigens prepared by acetone precipitation of culture-media supernatants. Strains of *A. israelii* produced two to five precipitin lines with heterologous *A. israelii* antiserum, but all strains had at least two bands in

**Figure 6-79. Interspecies cross-reactions demonstrable with FA.**

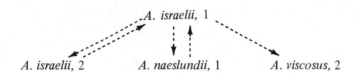

The cross-reactions between *A. israelii*, 1, *A. israelii*, 2 and *A. naeslundii*, 1 are reciprocal but weak and strain variable. About one-third of our *A. viscosus*, 2 cultures will fluoresce with *A. israelii*, 1 antiserum.

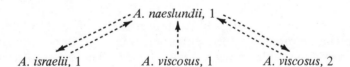

The cross-reactions between *A. naeslundii*, 1 and *A. viscosus*, 2 are quite complex, and some strains appear to have a very close antigenic relationship. *A. naeslundii* does not stain *A. viscosus*, 1, but *A. viscosus*, 1 antiserum stains *A. naeslundii* cells. The reaction with *A. israelii*, 1 is reciprocal.

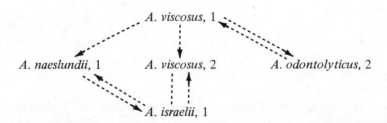

*A. viscosus*, 1 antiserum has a one-way cross with *A. naeslundii*, 1 and *A. viscosus*, 2 and a two-way cross with *A. odontolyticus*, 2. *A. viscosus*, 2 exhibits crosses with *A. viscosus*, 1, *A. naeslundii* and *A. israelii*, 1.

common. *A. bovis* was serologically distinct from *A. israelii,* but one line was formed by *A. naeslundii* antigen with *A. bovis* antisera. *A. israelii* antigens did not react with *A. naeslundii* antiserum, but again *A. naeslundii* antigens formed one line with *A. israelii* serum. Georg et al. (143) obtained similar results but did not demonstrate a cross-reaction between *A. naeslundii* and *A. bovis.* They distinguished between *A. bovis* smooth-colony serotype 1 and rough-colony serotype 2 but indicated that a cross-reaction occurred between serotype 1 antigen and serotype 2 antiserum. Later, Lambert et al. (264) found that only one of a large number of strains of *A. naeslundii* produced lines with *A. israelii* antisera.

Landfried (265) compared culture supernatant antigens with soluble antigen preparations made by the sonic disruption of washed cell suspensions. She found more precipitin lines, both species-specific and cross-reacting, in ID tests with the sonicated antigen preparations than with acetone-precipitated culture supernate antigens. With both types of antigens, species-specific precipitation patterns could be recognized, but there were more interspecies cross-reactions than in FA tests done with FITC conjugates of the same antiserum. *A. israelii,* *A. naeslundii* and *A. viscosus* cross-reacted as would be expected from FA studies. In addition, *A. bovis* also had reciprocal cross-reactions with the other three species. Antigenic differences were shown among stains of *A. israelii,* serotype 1. *A. odontolyticus* was not included in this study. Snyder et al. (435) also found a number of interspecies cross-reactions using sonically disrupted cells as antigens. Reciprocal reactions occurred between *A. israelii,* *A. naeslundii* and *A. odontolyticus* antigens and antisera. *A. bovis* produced one precipitin line with antisera for *A. israelii,* *A. naeslundii* and *A. odontolyticus,* but only *A. israelii* produced a precipitin line with the *A. bovis* antiserum. Reed and Gordon (384), studying *A. viscosus* and *A. naeslundii,* found that precipitation of culture supernatants with $NH_4SO_4$ gave more reproducible results than did precipitation with acetone, although essentially the same antigens were found in the two types of preparations. In this study, antigens prepared from mechanically disrupted cells did not produce more lines than did culture medium antigens. Five distinct bands of precipitation were seen with ID and immunoelectrophoresis. Serotype specific antisera could be produced by sorption.

It should be emphasized that there are two major problems with the ID tests as used to date: 1) the antigens are not standardized nor is the concentration controlled and 2) the antibody content of the antiserum is variable.

When antigens are precipitated from culture supernate, there is no way to standardize the preparations since they contain protein from the medium. The actual antigenic concentration may be variable even in similar preparations from the same cells. Perhaps of even greater importance is that the number of precipitin lines formed in any one system is dependent upon the particular lot of antiserum being used (42, 265, 266). This is exemplified by our experience with *A. viscosus* antisera from two different rabbits immunized with different schedules and tested with the same acetone-precipitated culture supernate. One antiserum produced four precipitin lines and also two cross-reacting lines with *A. naeslundii.* The other antiserum showed only two lines with the homologous antigen and no cross-reactions. In our experience, continued immunization of rabbits, while having little effect on the FA titer, increased the number of lines formed in the ID tests. Thus, for the ID test, *rabbits should be immunized until no additional lines are produced with the antigen.* These factors should be considered when laboratories report different results with the ID tests although using the same type of antigens.

**Cell-Wall Agglutination.** Cummins (9, 89) has studied four *Actinomyces* species using cell-wall agglutination tests. The tests use antisera made against partially disintegrated cell suspensions and purified cell walls as antigens. The results of CWA tests are shown in Table 6-13. The species can be separated by this technique as with FA and ID. *A. bovis* was antigenically distinct, while *A. israelii,* *A. naeslundii* and *A. viscosus* show low-level cross-reactions. High-titered (1:200) cross-reactions occurred with *A. naeslundii* cell walls and both *A. israelii,* 1 and *A. viscosus,* 2 antisera.

**Cross-Reactions with Other Genera.** *Actinomyces* show few cross-reactions with other genera, and when they occur they are usually low titered, strain variable and frequently show only a one-way cross. *A. israelii* may cross with *P. acnes; A. viscosus* with *Arachnia* and *Rothia;* and *A. odontolyticus* and *C. pyogenes* may exhibit reciprocal cross-reactions. In general, more inter-

Table 6-13. Results from cell-wall agglutination tests[a].

| Cell-Wall Antigens | A. israelii, 1 (ATCC 10048) | | A. naeslundii (ATCC 12104) | | A. viscosus, 2 (ATCC 19246) | | A. bovis, 1 (ATCC 13683) | | A. bovis, 1 (A1) | |
|---|---|---|---|---|---|---|---|---|---|---|
| | 20[b] | 200 | 20 | 200 | 20 | 200 | 20 | 200 | 20 | 200 |
| A. israelii, 1 (10048)[c] | ++ | ++ | − | − | ++ | − | − | − | ± | − |
| A. naeslundii, 1 (12104) | ++ | + | ++ | ++ | ++ | + | − | − | | |
| A. viscosus, 1 (19246) | ++ | − | ++ | − | ++ | − | − | − | − | − |
| A. viscosus, 2 (19246) | ++ | − | ++ | − | ++ | ++ | + | − | − | − |
| A. bovis, 1 (13683) | − | − | − | − | ++ | − | ++ | + | + | − |

[a] Data from Subgroup report (9).
[b] Dilutions of antisera used were 1:20 and 1:200.
[c] ATCC numbers of the cultures used.

generic cross-reactions have been shown with ID than with FA, but even so they are minimal. The widest range of cross-reactions has been reported by Kwapinski (261), who showed serological relationship between cytoplasmic antigens of *Actinomyces, Mycobacterium, Corynebacterium* and *Nocardia.*

**Antigens of *Actinomyces*.** Despite the wide use of serology for identification of *Actinomyces,* very little is known about the number, chemical composition or location in the cell of antigens involved in FA, ID or CWA reactions.

Cummins (89) attributed the CWA of corynebacteria, mycobacteria and nocardiae to a polysaccharide antigen and as *Actinomyces* antigens were prepared in the same way suggested they were similar in nature. King and Meyer (244) indicated that the acetone-precipitated antigens from culture supernate were extracellular polysaccharides from the cell wall. However, Kwapinski (259, 261) found that polysaccharide fractions showed limited serological activity and indicated that nucleoprotein fractions had a wider serological activity and that purified lipids were serologically inactive.

Pirtle et al. (373) isolated two water-soluble, heat-stable antigens from the supernatant broth cultures of *A. israelii* (ATCC 10048). The two differed in the percentage of hexose and nitrogen, but both had mannose as their chief component. More recent studies (42, 266) suggest that the mannose was probably derived from the yeast extract in the culture medium. Bowden and Hardie (42) studied antigens from all *Actinomyces* species except *A. bovis* and reported two groups of antigens. Group 1 antigens were cell-wall-associated carbohydrates and demonstrable by CWA or complement fixation but not by precipitin tests. They were resistant to trypsin and pronase and contained 34-62% carbohydrate, 1-9% amino sugars and 0.1-0.5 µg/mg amino acids. They were generally species-specific. The only significant interspecies cross-reaction was between *A. naeslundii* walls and *A. viscosus,* 2 antiserum. The group 2 antigens were demonstrable in both acid extracts of whole cells and in culture supernatant by precipitin or ID tests. They were resistant to trypsin but not pronase, indicating that an intact polypeptide was involved. These antigens were also generally species-specific, but cross-reactions occurred between *A. israelii,*

*A. naeslundii* and *A. viscosus.* Cross-reactions were also demonstrable between *A. israelii* serotypes 1 and 2 and between *A. viscosus* serotypes 1 and 2; however, these were minor, and the antigenic identity of the respective serotypes was demonstrable. The relationship of these group 1 and group 2 antigens to antigens demonstrable in FA tests is not known.

Landfried (266) isolated and characterized an antigen from the culture supernate of *A. israelii* (ATCC 12102). This antigen was nonmigratory in immunoelectrophoresis and present in all tested strains of both serotypes. The antigen was characterized as a glycan, probably of galactose, mixed with one or two other glycans, one of which was also antigenic. Reed (382) studied the cell-wall antigens of *A. viscosus* (ATCC 15987) and demonstrated that the peptidoglycan was the major antigen and that rhamnose was not a determinant.

In summary, polysaccharides, polypeptides, nucleopeptides and glycans have been indicated as antigenic determinants of the *Actinomyces,* but other components may also be important. Thus, at the present time, the antigenic makeup of the *Actinomyces* has not been adequately defined, and this is a fertile but difficult field for investigation.

# 7

# Genus *Arachnia*

## HISTORY

In 1962, Buchanan and Pine (56) proposed that a propionic-acid-producing actinomycete previously identified as *A. israelii* (367, 368) be classified as a new species and proposed the name *Actinomyces propionicus*. Further isolates were not recognized until 1967, when three additional strains of *A. propionicus* were described by Gerencser and Slack (145). The classification of this species was reviewed by Pine and Georg (364, 365), and they reclassified the organism as *Arachnia* (spiderlike) *propionica*.

## A. MORPHOLOGY

This organism is gram positive, growing as diphtheroidal cells (0.2 to 0.3 μm x 3 to 5 μm) and branching filaments which may be 5-20 μm in length (Figures 7-1, 7-2, 7-3). Swollen spherical cells are frequently seen, and an occasional culture may consist entirely of these coccoid forms (Figure 7-4). Buchanan and Pine (56) found that growth conditions influenced morphology.

Figure 7-1.  *A. propionica,* gram stain, Thioglycollate Broth, 48 hr., diphtheroidal cells, X1200.

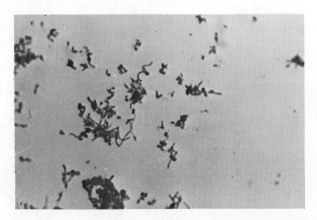

Figure 7-2.  *A. propionica,* gram stain, Thioglycollate Broth, 48 hr., diphtheroids and filaments, X1200.

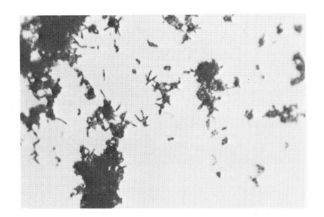

**Figure 7-3.** *A. propionica,* gram stain, Thioglycollate Broth, 48 hr., diphtheroids, filaments and coccoid forms, X1200.

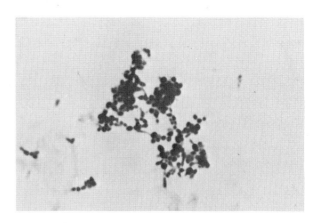

**Figure 7-4.** *A. propionica,* gram stain, Thioglycollate Broth, 48 hr., coccoid forms, X1200.

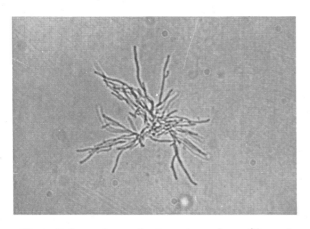

**Figure 7-5.** *A. propionica,* microcolony, filaments radiating from a central point. BHIA, 24 hr., X720.

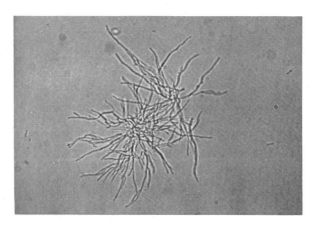

**Figure 7-6.** *A. propionica,* large, filamentous microcolony. BHIA, 24 hr., X720.

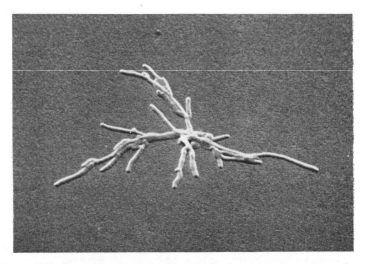

**Figure 7-7.** Scanning electron micrograph, *A. propionica,* microcolony, BHIA, 24 hr., branching, filaments from a single point. X2000. (Prepared by P. Allender, Division of Infectious Diseases, WVU.)

 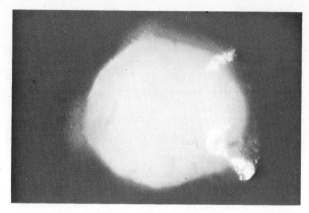

**Figure 7-8.** *A. propionica,* BHIA, 7 days, bread-
crumb colony, X40. (From Gerencser,
M. A., and Slack, J. M. J. Bacteriol.
*94:*109-115, 1967.)

**Figure 7-9** *A. propionica,* BHIA, 7 days, smooth
colony, X24.

Organisms grown in 1-liter volumetric flasks with pyrogallol + $N_2CO_3$ seals were long and thread-like and formed cottonlike masses. They observed the formation of swollen cells resembling spheroplasts which predominated when grown in broth with raffinose. These spherical cells resemble those in Figure 7-4, but in our experience these cells are formed in routine media and do not require special growth conditions. In thioglycollate broth, growth occurs as suspended discrete colonies, but, in BHI broth, the clumps settle out, leaving a clear supernate.

*Colony Morphology.* In 24 hrs. on BHI plates, *A. propionica* produces branching, filamentous microcolonies (Figures 7-5, 7-6, 7-7) closely resembling those of *A. israelii.* Strains producing smooth microcolonies are rare. Mature colonies (7-14 days) are more variable (Figures 7-8 and 7-9). Some isolates produce rough colonies of the molar-tooth or bread-crumb type resembling those of *A. israelii* while others produce smooth, convex colonies which often have undulate edges.

## B. OXYGEN REQUIREMENTS

*A. propionica* is a facultative anaerobe. It does grow aerobically, but it grows best anaero-bically. Under neither condition does it require nor is it stimulated by $CO_2$. Good growth is obtained earlier and better on anaerobic slants than on aerobic slants. Buchanan and Pine (56) found that it grew well aerobically in liquid medium, but anaerobic growth could be initiated by a smaller inoculum.

## C. BIOCHEMICAL REACTIONS

These reactions are shown in Table 7-1. The fourteen strains studied here and by Dr. L. Georg are quite uniform in their reactions. They show variability in M.R. test and with a few sugars, namely: glycerol, inositol, salicin and starch. *A. propionica* resembles *A. israelii* biochemically and morphologically but differs in that it does not hydrolyze esculin or ferment either arabinose or xylose but does ferment adonitol. Under anaerobic conditions, with or without $CO_2$, it fer-ments glucose to produce mainly acetic and propionic acids and $CO_2$ with small amounts of lac-tate or succinate (56). All strains so far reported (50, 145) produce propionic acid from glucose.

## D. SEROLOGY

Two serotypes of *A. propionica* have been recognized using FA techniques. The type strain (ATCC 14157) and two other strains studied by Gerencser and Slack (145) were closely related

**Table 7-1. Biochemical reactions of *Arachnia propionica*[a].**

| Test | Strains Positive[b] | %+ | Test | Strains Positive | %+ |
|---|---|---|---|---|---|
| Catalase | 0/14 | 0 | Adonitol[f] | 10/10 | 100 |
| Indole | 0/14 | 0 | Amygdalin | --------- | ------ |
| Methyl Red | 6/10 | 60 | Arabinose | 0/14 | 0 |
| Voges-Proskauer | 0/10 | 0 | Cellobiose | 0/8 | 0 |
| Nitrate Reduction | 14/14 | 100 | Dextrin | 0/1 | ------ |
| Nitrite Reduction | ------[e] | ----- | Dulcitol | 0/10 | 0 |
| Esculin Hydrolysis | 0/9 | 0 | Erythritol | 0/1 | ------ |
| Starch Hydrolysis | 8/8 | 100 | Fructose | 4/4 | 100 |
| Hippurate Hydrolysis | ------ | ----- | Galactose | 14/14 | 100 |
| Urease | 0/11 | 0 | Glucose | 14/14 | 100 |
| $H_2S$–BHI[c] | 0/8 | 0 | Glycerol | 4/14 | 29 |
| TSI | 6/8 | 75 | Glycogen | 0/10 | 0 |
| Gelatin Liquefaction | 0/8 | 0 | Inositol | 4/13 | 31 |
| Serum Liquefaction | ------ | ----- | Inulin | 0/10 | 0 |
| Casein Hydrolysis | ------ | ----- | Lactose | 14/14 | 100 |
| Milk-A-AR[d] | 2/8 | 25 | Maltose | 14/14 | 100 |
| AC | 4/8 | 50 | Mannitol | 14/14 | 100 |
| P | 0/8 | 0 | Mannose | 4/4 | 100 |
| NC | 2/8 | 25 | Melezitose | 0/4 | 0 |
| | | | Melibiose | 0/1 | ------ |
| | | | Raffinose | 8/8 | 100 |
| | | | Rhamnose | 0/14 | 0 |
| | | | Ribose | 0/1 | ------ |
| | | | Salicin | 2/14 | 14 |
| | | | Sorbitol | 10/10 | 100 |
| | | | Starch | 11/14 | 79 |
| | | | Sucrose | 14/14 | 100 |
| | | | Trehalose | 14/14 | 100 |
| | | | Xylose | 0/14 | 0 |

[a] Data from Pine and Georg (366) and WVU unpublished data.
[b] Number of strains positive/number of strains tested.
[c] BHI = Brain Heart Infusion; TSI = Triple Sugar Iron.
[d] A = acid; AR = acid, reduction; AC = acid, clot; P = peptonization
   NC = no change.
[e] ____ = no data available.
[f] Basal Medium–Thioglycollate Fermentation Base.

or identical, but the fourth strain used in this study (WVU 346) was serologically different. Brock et al. (50) studied nine additional cultures which also fell into two serotypes using strain ATCC 14157 as the prototype of serotype 1 and strain WVU 346 as the prototype of serotype 2. Some cross-staining occurred between the serotypes which could be removed by sorption. Holmberg and Forsum (206) prepared defined immunofluorescent reagents against these two strains. With a 3+ fluorescence endpoint, the two serotypes did not cross-react, but, when 1+ fluorescence was the endpoint, low-titered cross-staining occurred. Gel diffusion (145) studies also showed that the strains of serotype 1 were closely related and that cross-reactions occurred between serotypes. Serotype 2 strain 346 formed only one line with serotype 1 antiserum. In contrast, Johnson and Cummins (225) reported that *A. propionica* 5067 (WVU 346) did not react in cell-wall agglutination tests with antiserum to the type strain (ATCC 14157).

*A. propionica* shows no serological relationship to other actinomycetes.  FA reactions with species of *Actinomyces*, *Rothia*, *Bifidobacterium*, *Corynebacterium*, *Nocardia* and *Bacterionemia* are negative (50, 145, 206).  We have seen some cross-reactions between a freshly isolated strain of *P. acnes* and *Arachnia* antisera, but these cross-reactions have not been observed by other workers.

**Cell Wall.**  The composition of the cell wall of *A. propionica* has been discussed in the chapter on cell walls.  All strains studied of both serotypes contain LL-Dap as the major diamino acid.

**DNA.**  Johnson and Cummins (225) reported % GC for five strains of *A. propionica* ranging from 63 to 68%.  They had from 87 to 95% homology with reference DNA from the type strain.

**Classification.**  Questions have been raised about the nomenclatural place of this organism, which on the basis of morphology was first placed in the genus *Actinomyces* and then in a new genus *Arachnia* (spiderlike) which was created because the combination of forming a filamentous microcolony, producing propionic acid and having Dap in the cell wall, was unique.  More recently, the question has been raised as to whether or not it should be more rightfully included in *Propionibacterium* because of the propionic acid produced and the presence of Dap.  This warrants continued consideration, but, on the basis of its pathogenicity for man, production of sulfur granules, filamentous microcolony and lack of DNA homologies (225), it seems best to leave it in the separate genus *Arachnia*.

Some comment should be made concerning *A. propionica* WVU 346 (CDC, W904), which was so identified on the basis of its morphological and biochemical characteristics (50, 145).  It was serologically different and designated as serotype 2 (50) following which three additional type 2 isolates were identified (145).  Johnson and Cummins (225) studied the strain (VPI 5067) and concluded that it was not *A. propionica* because it showed no DNA homology with the type strain, although the % GC was the same.  They found L-Dap and no cross-reactions using the cell-wall agglutination test.  These results suggest that this and the three additional isolates may represent a new species of *Arachnia*, but further study of these and additional isolates is required to establish such a species.

# 8

# Genus *Rothia*

## HISTORY

Onisi (344) proposed the name *Actinomyces dentocariosus* for a group of catalase positive, facultative anaerobes isolated from carious dentine. The organisms were gram-positive diphtheroids or branching filaments which often fragmented into coccoid forms and produced both R and S colonies on solid media. Later, Onisi and Nuckolls (345) described further isolates from carious dentine which were morphologically and biochemically like *Actinomyces dentocariosus.* They were catalase positive, liquefied gelatin (16/22) and produced acid from glucose, maltose, sucrose and inulin. In this paper, the authors stated that these organisms, referred to as type III B, might be identified with the genus *Nocardia,* but no specific epithet was used, and the name *Actinomyces dentocariosus* was not mentioned.

Roth (393) described proteolytic organisms from carious lesions dividing them into five groups, one of which was identical with Onisi's *Actinomyces dentocariosus.* Roth suggested that the organism be renamed *Nocardia dentocariosus* because of its preference for aerobic growth. She also points out that the bacteria described by Morris (325) as group 7 were probably *N. dentocariosus.* Roth and Thurn (394) described additional strains of *N. dentocariosus* in greater detail. During this same time, Davis and Freer (95) in England described twenty-five strains of an aerobic actinomycete from the human mouth. They proposed the name *Nocardia salivae* despite the fact that the cell walls contained lysine rather than Dap. Georg and Brown (139) compared strains of *N. dentocariosus* and *N. salivae* and concluded that they were identical. Since neither *Actinomyces* nor *Nocardia* was an appropriate genus for these organisms, a new genus *Rothia* was created.

## A. CELL MORPHOLOGY

Gram-stained smears from broth cultures show pleomorphic, gram-positive organisms which range from coccoid to filamentous forms (Figures 8-1, 8-2, 8-3). Brown et al. (52) reported that most of their isolates were completely coccoid, while others had coccoid and bacillary forms. Only an occasional culture was completely filamentous. In our experience, most cultures are mixtures of diphtheroidal, filamentous and coccoidal forms. At times, a culture would be predominantly coccoid or predominantly filamentous, but we have not been able to obtain and maintain completely coccoid or completely filamentous cultures of a strain. Loss of gram-positivity is seen in some filaments (52) and in aging cultures (282). Agar-grown cultures usually show a

71

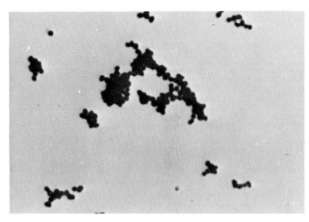

Figure 8-1. *R. dentocariosa,* gram stain, coccoid form, TSB, 48 hr., X1600. (From Brown et al. Appl. Microbiol. *17:* 150-156, 1969.)

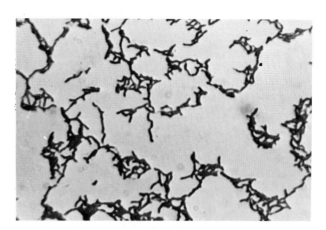

Figure 8-2. *R. dentocariosa,* gram stain, rods and filaments, TSB, 48 hr., X700. (From Brown et al. Appl. Microbiol. *17:*150-156, 1969.)

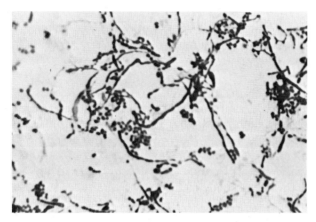

Figure 8-3. *R. dentocariosa,* gram stain, rods, filaments and coccoid forms, TSB, 48 hr., X700. (From Brown et al. Appl. Microbiol. *17:*150-156, 1969.)

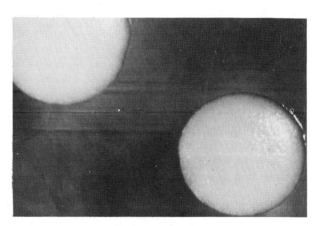

Figure 8-4. *R. dentocariosa,* smooth colony, 48 hr., aerobic TSA plate, X5. (From Brown et al. Appl. Microbiol. *17:*150-156, 1969.)

predominance of rods and branched filaments. Coccoid forms varied from 1 $\mu$m-5 $\mu$m in diameter. Filamentous forms are approximately 1 $\mu$m in width and do not exceed 35 $\mu$m in length (282).

**Colony Morphology.** Young (18-24 hr.) colonies on aerobic TSA plates average 1.0 mm in diameter. When observed microscopically, most colonies have an entire edge with a smooth or granular surface, and only a small percentage have a filamentous edge (Figure 8-4). In contrast, young colonies on *anaerobic* TSA plates generally have filamentous edges and, although larger, resemble filamentous colonies of *Actinomyces* (Figure 8-5). These anaerobically grown colonies also average 1 mm in diameter and have a granular center. Roth and Thurn (394) showed that young colonies growing under a coverslip had filamentous edges while other colonies on the same plate did not.

After 7 days, mature colonies are 1-4 mm in diameter and vary from smooth to very rough. Most strains have colonies which are raised to convex with undulate or scalloped edges and highly convoluted surfaces (Figures 8-6, 8-7, 8-9). Other strains produce colonies which are convex, with entire edges and smooth surfaces (Figure 8-8). Often, both types of colonies are produced by a given strain, and mixtures of S and R colonies may be observed on the same plate. The colonies vary in texture from mucoid to bread-crumb-like and are creamy white.

**Cell-Wall Composition.** This is discussed in Chapter 5 on cell walls. Like *Actinomyces,* *Rothia* contains lysine as the major diamino acid and does not contain arabinose.

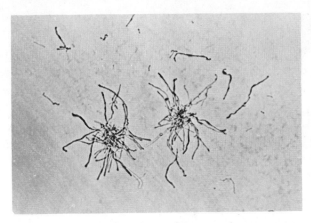

Figure 8-5. *R. dentocariosa*, "spider" colony, 48 hr., anaerobic TSA plate, X310. (From Brown et al. Appl. Microbiol. *17:*150-156, 1969.)

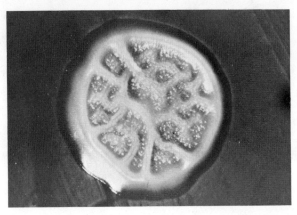

Figure 8-6. *R. dentocariosa*, rough surface with entire edge, aerobic TSA plate, 7 days, X10. (From Brown et al. Appl. Microbiol. *17:*150-156, 1969.)

Figure 8-7. *R. dentocariosa*, rough, scalloped edge, aerobic TSA plate, 7 days, X5. (From Brown et al. Appl. Microbiol. *17:*150-156, 1969.)

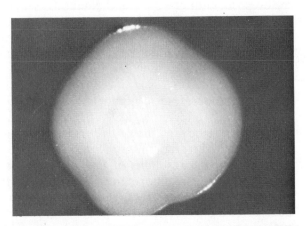

Figure 8-8. *R. dentocariosa*, smooth colony, aerobic TSA plate, 7 days, X25.

## B. OXYGEN REQUIREMENTS

Onisi (344) and Onisi and Nuckolls (345) described their isolates as microaerophilic or facultative, while Roth described *N. dentocariosus* as aerobic to facultative with a preference for aerobic growth. Using BHIA slants with various seals, Brown et al. (52) found that all strains grew best aerobically, except one, and that $CO_2$ seemed to reduce growth. Growth under anaerobic + $CO_2$ seals is poor or absent. In this laboratory, Lesher found that all cultures grew aerobically and that the addition of $CO_2$ seemed to reduce growth. He also demonstrated that stationary aerobic cultures in TSB had a 1-3 hr. lag phase with maximum exponential growth at 12-15 hrs.

Despite the preference for aerobic growth in $O_2$ requirement tests, all strains of *Rothia* do grow in an anaerobe jar with $N_2:H_2:CO_2(80:10:10)$, and many of our strains were isolated from anaerobic plates. The ability of these cultures to grow in prereduced media inoculated under $O_2$ free gas or in an anaerobic chamber has not been tested.

## C. BIOCHEMICAL REACTIONS

*Rothia* is catalase positive, although anaerobically grown cultures should be exposed to air for 30 min. before being tested. Biochemical reactions have been reported by a number of

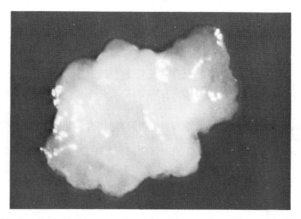

**Figure 8-9.** *R. dentocariosa,* **bread-crumb colony, aerobic**
**TSA plate, 7 days, X25.**

authors, including: Onisi and Nuckolls (345), Roth (393), Roth and Thurn (394), Davis and Freer (95), Howell and Jordan (211), Brown et al. (52) and Holmberg and Hallender (207). There is good agreement with the results listed in Table 8-1, except several authors (95, 211, 207) report positive V-P tests. Roth and Thurn (394) indicated *Rothia* was proteolytic because of its action on hide powder, but others report negative or variable results. Gelatin hydrolysis is demonstrable on a gelatin agar plate but is usually negative in the tube test.

In this laboratory, a number of strains which resembled *Rothia* morphologically but differed biochemically and serologically from the type strain ATCC 17931 have been isolated from dental calculus. A study of fifty of these strains, including typical *Rothia* and *Rothia*-like cultures (282), showed that the strains were *Rothia* but could be divided into four biotypes and three serotypes. The biochemical reactions of the four biotypes are shown in Table 8-2. Biotype 1 strains were uniform in their reactions and conformed to the previous description of *R. dentocariosa*. Biotypes 2 and 3 showed increasing degrees of variation, and all strains differed from biotype 1 in at least two characteristics. Biotype 2 strains varied in nitrite reduction (57%) and showed significant variation in the number of strains which fermented lactose (20%), mannose (57%), ribose (14%), salicin (28%) and trehalose (57%). Biotype 3 strains were all urease positive, gave variable nitrite reduction (63%) and also showed variation in the fermentation of lactose (50%), ribose (29%), salicin (71%) and trehalose (71%). Biotype 4 strains showed the greatest degree of difference and were consistently more active in the fermentation of mannitol, raffinose and lactose. In all biotypes, a strain which varied in its sugar fermentations usually differed in several sugars so that a strain which failed to ferment mannose usually also failed to ferment ribose, salicin and trehalose. In the same way, strains which fermented lactose often also fermented rhamnose.

## D. SEROLOGY

Ritz (386) immunized rabbits with a mixture of six oral strains of *Nocardia (Rothia)* and then using indirect FA identified known strains and twenty-nine oral isolates as *R. dentocariosa*. He also localized *Rothia* in the surface layers of sections of dental plaque. Weak cross-reactions were seen with some strains of *Actinomyces, Lactobacillus, Bacterionema, N. asteroides* and *N. brasiliensis*. Recently, Holmberg and Forsum (206) prepared FITC conjugate for *R. dentocariosa* (ATCC 17931) from Fab fragments of IgG to avoid possible cross-reactions due to staphylococcal protein A. They obtained staining of all ten *Rothia* strains used in their numerical taxonomy study.

Snyder et al. (436) showed a close relationship between *N. dentocariosa* and *N. salivae* using the agglutination and gel diffusion tests. Hammond (181) isolated a soluble polysaccharide (RPS) antigen from the cell wall of *R. dentocariosa* (ATCC 17931). The major sugars of RPS were fructose, glucose, galactose and ribose, with fructose the major antigenic determinant. The RPS

**Table 8-1. Biochemical reactions of *Rothia dentocariosa*.**

| Test | Brown et al. (52) 50 strains | | Lesher et al. (282)[a] 39 strains | | Total 89 strains | |
|------|------|------|------|------|------|------|
|  | No.+ | %+ | No.+ | %+ | No.+ | %+ |
| Catalase | 50 | 100 | 39 | 100 | 89 | 100 |
| Indole | 0 | 0 | 0 | 0 | 0 | 0 |
| Nitrate | 50 | 100 | 37 | 95 | 87 | 98 |
| Nitrite - 0.01% | 50 | 100 | 29 | 74 | 79 | 89 |
| 0.001% | -[d] | - | 30/32 | 94 | 30/32 | 94 |
| Esculin | 50 | 100 | 38 | 97 | 88 | 99 |
| Hippurate Hydrolysis | - | - | 0 | 0 | 0 | 0 |
| Urease | 0 | 0 | 7 | 18 | 7 | 8 |
| H$_2$S - HIA[b] | 1 | 2 | - | - | 1/50 | 2 |
| TSI | 48 | 96 | - | - | 48/50 | 96 |
| Gelatin Liquefaction | 0 | 0 | 0 | 0 | 0 | 0 |
| Serum Liquefaction | - | - | 0 | 0 | 0 | 0 |
| Casein Hydrolysis | - | - | 0 | 0 | 0 | 0 |
| Milk | | | | | | |
| No change | 50 | 100 | - | - | 50/50 | 100 |
| Adonitol[c] | 0 | 0 | 0 | 0 | 0 | 0 |
| Arabinose | 0 | 0 | 0 | 0 | 0 | 0 |
| Cellobiose | 0 | 0 | 0 | 0 | 0 | 0 |
| Glucose | 50 | 100 | 39 | 100 | 89 | 100 |
| Glycerol | 36 | 72 | 39 | 100 | 75 | 84 |
| Glycogen | - | - | 0 | 0 | 0 | 0 |
| Inositol | 0 | 0 | 0 | 0 | 0 | 0 |
| Lactose | 0 | 0 | 6 | 15 | 6 | 7 |
| Maltose | 50 | 100 | 39 | 100 | 89 | 100 |
| Mannitol | 0 | 0 | 1 | 3 | 1 | 1 |
| Mannose | - | - | 35 | 90 | 35/39 | 90 |
| Raffinose | 0 | 0 | 1 | 3 | 1 | 1 |
| Rhamnose | - | - | 4 | 10 | 4/39 | 10 |
| Ribose | - | - | 25 | 64 | 25/39 | 64 |
| Salicin | 50 | 100 | 32 | 82 | 82 | 92 |
| Sorbitol | - | - | 0 | 0 | 0 | 0 |
| Starch | 0 | 0 | 0 | 0 | 0 | 0 |
| Sucrose | 50 | 100 | 39 | 100 | 89 | 100 |
| Trehalose | - | - | 34 | 87 | 34/39 | 87 |
| Xylose | 0 | 0 | 0 | 0 | 0 | 0 |

[a] See Table 8-2 for separation of biotypes.
[b] HIA = Heart Infusion Agar; TSI = Triple Sugar Iron Agar.
[c] Basal Medium = Meat extract + peptone with Andrade's indicator.
[d] Not done.

## Table 8-2. Biochemical reactions of *Rothia* biotypes[a].

| Test | Biotype 1 25 strains No.+ | %+ | Biotype 2 7 strains No.+ | %+ | Biotype 3 7 strains No.+ | %+ | Biotype 4 11 strains No.+ | %+ |
|---|---|---|---|---|---|---|---|---|
| Catalase | 25 | 100 | 7 | 100 | 7 | 100 | 11 | 100 |
| Indole | 0 | 0 | 0 | 0 | 0 | 0 | 0 | 0 |
| Nitrate | 25 | 100 | 6 | 86 | 6 | 86 | 11 | 100 |
| Nitrite - 0.01% | 21 | 84 | 4 | 57 | 4 | 57 | 4 | 36 |
| 0.001% | 25 | 100 | —[d] | — | 5 | 71 | 8 | 72 |
| Esculin | 25 | 100 | 7 | 100 | 6 | 86 | 9 | 82 |
| Hippurate | 0 | 0 | 0 | 0 | 0 | 0 | 0 | 0 |
| Urease | 0 | 0 | 0 | 0 | 7 | 100 | 3 | 27 |
| $H_2S$ - HIA[b] | — | — | — | — | — | — | — | — |
| TSI | — | — | — | — | — | — | — | — |
| Gelatin-Thiogel | 0 | 0 | 0 | 0 | 0 | 0 | 0 | 0 |
| Loeffler's Serum | 0 | 0 | 0 | 0 | 0 | 0 | 0 | 0 |
| Casein Hydrolysis | 0 | 0 | 0 | 0 | 0 | 0 | 0 | 0 |
| Milk | | | | | | | | |
| No change | Not tested | | | | | | | |
| Adonitol[c] | 0 | 0 | 0 | 0 | 0 | 0 | 0 | 0 |
| Arabinose | 0 | 0 | 0 | 0 | 0 | 0 | 0 | 0 |
| Cellobiose | 0 | 0 | 0 | 0 | 0 | 0 | 0 | 0 |
| Glucose | 25 | 100 | 7 | 100 | 7 | 100 | 11 | 100 |
| Glycerol | 25 | 100 | 7 | 100 | 7 | 100 | 11 | 100 |
| Glycogen | 0 | 0 | 0 | 0 | 0 | 0 | 0 | 0 |
| Inositol | 0 | 0 | 0 | 0 | 0 | 0 | 0 | 0 |
| Lactose | 0 | 0 | 2 | 28 | 4 | 51 | 5 | 45 |
| Maltose | 25 | 100 | 7 | 100 | 7 | 100 | 11 | 100 |
| Mannitol | 0 | 0 | 0 | 0 | 1 | 14 | 9 | 82 |
| Mannose | 25 | 100 | 4 | 57 | 6 | 86 | 11 | 100 |
| Raffinose | 0 | 0 | 0 | 0 | 1 | 14 | 9 | 82 |
| Rhamnose | 1 | 4 | 1 | 14 | 2 | 29 | 10 | 91 |
| Ribose | 22 | 88 | 1 | 14 | 2 | 29 | 10 | 91 |
| Salicin | 25 | 100 | 2 | 29 | 5 | 71 | 10 | 91 |
| Sorbitol | 0 | 0 | 0 | 0 | 0 | 0 | 0 | 0 |
| Starch | 0 | 0 | 0 | 0 | 0 | 0 | 0 | 0 |
| Sucrose | 25 | 100 | 7 | 100 | 7 | 100 | 11 | 100 |
| Trehalose | 25 | 100 | 4 | 57 | 5 | 71 | 10 | 91 |
| Xylose | 0 | 0 | 0 | 0 | 0 | 0 | 0 | 0 |

[a]Biotypes 1, 2, 3 considered *R. dentocariosa* (see Table 8-1); identity of biotype 4 uncertain. Data from Lesher et al. (282).
[b]HIA = Heart Infusion Agar; TSI = Triple Sugar Iron Agar.
[c] Basal Medium = Meat extract + peptone with Andrade's indicator.
[d] Not done.

was a common and specific antigen in all twenty-six strains and could be demonstrated by FA and gel diffusion techniques.

These reports suggest that *R. dentocariosa* is serologically homogeneous, and we found that this was true until we started using a *R. dentocariosa* 17931 conjugated antiserum to directly identify *Rothia* in dental material and isolates from plaque and calculus. Then it became obvious' that organisms morphologically and biochemically resembling *Rothia* were not staining, so conjugated antiserum was prepared against four of these strains (WVU 477, 936, 999 and 1088). The results of cross-staining and reciprocal sorption studies are shown in Table 8-3. These results tentatively establish three serotypes, and these antisera have been used in an attempt to serotype fifty isolates. Ten isolates typed as serotype 1, 15 as 2, and 9 as 3, but sixteen were either negative or equivocal. Thus, much more work has to be done to establish these or additional serotypes. The distribution of the four biotypes among these serotypes is shown in Table 8-4.

**Classification.** The various names which have been used for *Rothia* were discussed in the section on history. Georg et al. (139) detailed the reasons for not including these organisms in either *Actinomyces* or *Nocardia*. *Nocardia* was considered inappropriate because *Rothia* ferments glucose, requires an organic N source and does not contain either Dap or arabinose in its cell walls. *Actinomyces* was considered inappropriate because *Rothia* is more pleomorphic, producing filamentous to completely coccoid forms, is aerobic and produces lactic acid as a major product of glucose fermentation with little or no succinic acid. In general, *Rothia* is more like *Actinomyces* than *Nocardia*.

The genus contains only one species, and all of the strains described earlier were quite uniform in their characteristics. As additional strains have been studied, it has become apparent that there are biochemical as well as serological differences. Therefore, a major consideration is whether or not these strains should be designated as new species. The strains designated as biotypes 1, 2 and

### Table 8-3. Serological reactions of *Rothia* using FA.

| Antiserum | Sorbed with | Strain | | | | |
| | | 477 | 1088 | 1489 | 936 | 999 |
|---|---|---|---|---|---|---|
| 1489[b] | --- | - | - | + | - | - |
| | 477 | - | - | + | - | - |
| | 1088 | - | - | + | - | - |
| 477 | --- | + | + | + | - | - |
| | 1088 | + | - | - | - | - |
| | 1489 | + | - | - | - | - |
| 1088 | --- | + | + | - | - | - |
| | 477 | - | - | - | - | - |
| | 1489 | + | + | - | - | - |
| 936 | --- | - | - | - | + | + |
| 999 | --- | - | - | - | + | + |
| | 936 | - | - | - | - | - |

[a] + = 2+ or greater fluorescence at working titer (unsorbed sera) or at 1:2 (sorbed sera).

[a] - = less than 2+ fluorescence undiluted (unsorbed) or 1:2 (sorbed).

[b] Tentatively these have been indicated as serotypes 1, 2 and 3. Serotype 1 is represented by 1489; serotype 2 by 477 and 1088 and serotype 3 by 936 and 999.

**Table 8-4.  Relationship between *Rothia* biotypes and serotypes.**

| Serotypes | Biotypes | | | |
|---|---|---|---|---|
| | 1 | 2 | 3 | 4 |
| 1[a] | 6 | 2 | 2 | 0 |
| 2 | 9 | 3 | 3 | 0 |
| 1 & 2 | 7 | 1 | 2 | 0 |
| 3 | 0 | 0 | 0 | 9 |
| Neg | 3 | 1 | 0 | 2 |

[a] Serotype 1 is 1489 antiserum; 2 is 477 antiserum; 3 is 936 antiserum.

3 (Table 8-2) do differ among themselves both biochemically and serologically, although it is doubtful that this warrants a separate species name.  On the other hand, biotype 4 with its greater degree of difference should probably be designated as a separate species, although additional biochemical tests should be done as well as a more detailed antigenic analysis and DNA base ratios and homologies prior to making this decision.

# 9

# Genus *Bacterionema*

## INTRODUCTION AND HISTORY

The genus *Bacterionema* (thread-shaped rod) with the species *B. matruchotii* (Matruchot, a French mycologist) was created by Gilmour, Howell and Bibby (157) in order to separate it from *Leptotrichia buccalis*. These two organisms had never been satisfactorily separated so these authors recommended that the aerobic organism with branching filaments be named *B. matruchotii* and the anaerobic organism with nonbranching filaments be retained as *L. buccalis*. The early history and the problems of nomenclature with *Bacterionema* are reviewed by Gilmour et al. (157) and Davis and Baird-Parker (93).

## A. MORPHOLOGY

This organism is gram positive, but gram-negative filaments with gram-positive granules frequently occur in older cultures. A unique morphological characteristic is the presence of a long, nonseptate filament with an attached "bacillus body" at one end (Figures 9-1 and 9-2). These

Figure 9-1. *B. matruchotii,* gram stain, filaments with "bacillus body," BHI (½ strength), 24 hr., X1200.

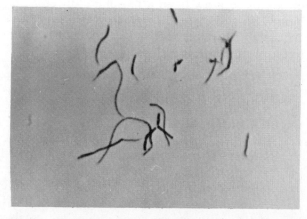

Figure 9-2. *B. matruchotii,* gram stain, filaments, shorter rods, coccoid forms, BHI (½strength), 24 hr., X1200.

filaments are 1-1.5 μm wide and 20-200 μm long, and the attached bacillus is 1-2.5 μm wide and 3-10 μm long. Septate filaments and dichotomous branching, which is most common in aerobic cultures, are also seen. Coccoid elements 0.4 μm in diameter may occasionally be observed. Reproduction is by fragmentation with subsequent elongation of the bacillary forms (155). The cells contain polymetaphosphate granules and lipid inclusions. Additional details are reported by Gilmour et al. (154, 157, 158), Baird-Parker and Davis (13) and Richardson and Schmidt (385).

Electron microscopy (449) has shown cell walls typical of gram-positive bacteria with an additional outer thin layer. This outer layer gives rise to membrane-limited vesicles extending out from the cell. Small spherical bodies, apparently free from the wall but membrane-bound, are seen along the cell. The filament walls are approximately 20 μm thick, and the bacillus walls 45-50 μm thick with an abrupt change in width between the two structures.

*Colony Morphology.* These descriptions are complicated by the fact that this organism will grow both aerobically and anaerobically with somewhat different morphologies.

*Young microcolonies,* aerobically or anaerobically grown, are flat and filamentous and resemble those of *A. israelii* (Figures 9-3 and 9-4). Aerobically grown strains may have a dense center with radiating filaments (Figure 9-5). The microcolonies contain septate, nonseptate and fragmenting filaments.

*Mature colonies* on aerobic plates are 0.5 to 1.5 mm in diameter and vary in appearance. Rough colonies may be circular, convex with a filamentous or entire margin (Figure 9-6), irregular with a convoluted surface and an entire or filamentous margin or irregular with a raised, wrinkled edge. All three types are opaque and adherent to the medium. Rough to smooth variation occurs frequently when cultures are maintained aerobically. Smooth colonies are circular to irregular, convex or umbonate with entire or lobate edges, opaque, soft and nonadherent. Anaerobically grown isolates are 1-2 mm, flat, filamentous, opaque at the center to translucent at the edge, tough and adherent. When R-to-S variation occurs, the mature colonies resemble R colonies on aerobic plates in having a narrow, filamentous edge and a convex or molar-tooth surface and are tough to soft and nonadherent.

Subsurface colonies in pour plates are fluffy and look like a ball of hair. In liquid media, *B. matruchotii* usually forms discrete masses of variable size. Smooth strains may produce a diffuse growth.

## B. OXYGEN REQUIREMENTS

*B. matruchotii* is a facultative anaerobe, although it has been variously classed as a strict aerobe, microaerophile or anaerobe (154, 155, 158, 215, 216). Strains maintained either aerobically or anaerobically tend to be considered either as strict aerobes or strict anaerobes, but most can be adapted to growth under either condition. The growth medium affects the ability of strains to grow in an atmosphere different from the one in which they were isolated. On non-blood-containing media, growth is best aerobically, but, on blood-containing media, aerobic and anaerobic growth is similar. Physiological studies (154, 215) show both fermentative and respiratory metabolism.

*DNA.* Several strains of *B. matruchotii* were studied by Page and Krywolap (351). The G + C base ratio, determined by TM, ranged from 55-58 moles %.

## C. BIOCHEMICAL REACTIONS

Biochemical reactions are shown in Table 9-1. The results are usually similar when tests are incubated aerobically or anaerobically, but differences which occur are shown in Table 9-2. *B. matruchotii* is benzidine positive, reduces nitrate to nitrite, hydrolyzes esculin, starch and hippurate and produces acetoin. The catalase test is usually positive under both aerobic and anaerobic conditions; however, some strains are catalase negative or doubtful when grown anaerobically but positive when grown aerobically. A few anaerobic strains remain catalase negative even when grown aerobically.

Figure 9-3. *B. matruchotii*, microcolony, "spider" type, BHIA, 24 hr., X166.

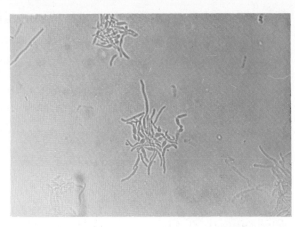

Figure 9-4. *B. matruchotii*, microcolony, BHIA, 24 hr., X720.

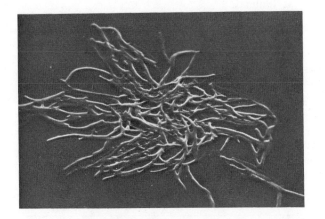

Figure 9-5. Scanning electron micrograph of *B. matruchotii*, microcolony, BHIA, 24 hr., showing dense center with long, branching filaments, 45° tilt, 20 KV, X1000. (Prepared by P. Allender, Division of Infectious Diseases, WVU.)

Figure 9-6. *B. matruchotii*, rough, wrinkled surface, undulate edge, BHIA, 10 days, X10.

*B. matruchotii* is indole, M.R., hydrogen-sulfide, lipase, acid-phosphatase and gelatin-hydrolysis negative. It produces no change in litmus milk and does not grow in 10% bile or 6.5% sodium chloride. About one-half the strains are weakly urease positive. It ferments carbohydrates with the production of acid and a small amount of $CO_2$. The gas production is not demonstrable in a Durham tube. Acid is produced from dextrin, fructose, glucose, maltose, mannose, salicin and sucrose. About 40% of the strains ferment raffinose, and an occasional strain ferments galactose, lactose, mannitol or trehalose.

The end products of glucose fermentation in stationary aerobic cultures are formic, acetic, propionic, lactic and succinic acids, acetyl-methyl carbinol and $CO_2$. If the cultures are grown on a shaker, succinic acid is not produced and lactic acid may or may not be present (154). Howell and Pine (215) found that, at least in some strains, the end products of glucose fermentation were dependent on oxygen availability. Very small amounts of volatile and nonvolatile acids were produced in shake cultures while formic, acetic, propionic, lactic and succinic acids and $CO_2$ were found in stationary cultures. They found that lactate was not used as a substrate in stationary cultures but was fermented in shaker-grown cultures and concluded that the accumulation of acids during stationary growth represented an accumulation of intermediates rather than the actual end products of glucose metabolism. More recently, Iwami et al. (223) studied lactate degradation by resting cells. Anaerobically very little $CO_2$ was formed, but aerobically lactate was con-

**Table 9-1. Biochemical reactions of *Bacterionema matruchotii*[a].**

| Test | No. Strains +/ No. Tested | % + | Test | No. Strains +/ No. Tested | % + |
|---|---|---|---|---|---|
| Catalase | 173/173 | 100 | Arabinose | 0/111 | 0 |
| Benzidine | 25/25 | 100 | Cellobiose | 0/55 | 0 |
| Indole | 0/147 | 0 | Dextrin | 52/56 | 93 |
| $NO_3 \rightarrow NO_2$ | 172/172 | 100 | Dulcitol | 0/67 | 0 |
| Methyl Red | 0/37 | 0 | Fructose | 55/55 | 100 |
| Voges-Proskauer[b] | 168/173 | 97 | Galactose | 6/136 | 4 |
| $H_2S$ | 0/176 | 0 | Glucose | 136/136 | 100 |
| Urease (weak) | 25/56 | 45 | Glycerol | 0/150 | 0 |
| Esculin Hydrolysis | 89/99 | 90 | Glycogen | 0/55 | 0 |
| $NH_4$ from arginine | 0/55 | 0 | Inositol | 0/12 | 0 |
| Hippurate Hydrolysis | 96/99 | 97 | Inulin | 0/111 | 0 |
| Starch Hydrolysis[b] | 67/167 | 40 | Lactose | 5/136 | 4 |
| Gelatin Hydrolysis | 0/147 | 0 | Maltose[b] | 108/111 | 97 |
| Litmus Milk | 0/48 | 0 | Mannitol | 7/67 | 10 |
| Lipase | 0/44 | 0 | Mannose[b] | 66/67 | 99 |
| Acid Phosphatase | 0/44 | 0 | Melezitose | 0/67 | 0 |
| 10% bile – growth | 0/36 | 0 | Raffinose[b] | 46/111 | 41 |
| 6.5% NaCl – growth | 0/36 | 0 | Salicin[b] | 63/67 | 94 |
| | | | Sorbitol | 0/67 | 0 |
| | | | Sucrose | 111/111 | 100 |
| | | | Trehalose | 3/92 | 3 |
| | | | Xylose | 0/67 | 0 |
| | | | Rhamnose | 0/67 | 0 |

[a] Data from Richardson and Schmidt (385); Howell and Pine (215); Gilmour and Beck (155); Takazoe et al. (448); Gilmour (154).

[b] Results same for aerobic and anaerobic tests except where marked b (see Table 9-2).

verted to acetate and $CO_2$. Iwami et al. (223) also concluded that conversion of lactate to other acids and $CO_2$ was dependent on the amount of oxygen available.

*B. matruchotii* has both fermentative and respiratory metabolism and is able to oxidize TCA-cycle acids (154). Smith and Strekfuss (432) were able to demonstrate a number of respiratory-linked dehydrogenases including succinate, maleate and NADH. These dehydrogenases apparently have different pathways to molecular oxygen. Oxidized cytochrome c was readily reduced in the presence of succinate and NADH. The authors suggest the presence of an active dicarboxylic acid cycle. Baboolal (11) studied the esterase, glucose-6-phosphate dehydrogenase and 6-phosphogluconate dehydrogenase isozyme patterns of *Bacterionema*, *Leptotrichia*, *Fusobacterium* and "anaerobic filaments" using starch-gel electrophoresis. The six strains of *B. matruchotii* formed a homogeneous group producing a distinctive pattern for each of the three enzymes which separated them from the other genera studied.

In nutritional studies, Gilmour and Bibby (156) were able to grow the organism in a synthetic medium containing glucose, eighteen amino acids, riboflavin, thiamine, nicotinic acid, pantothenic acid and cysteine. Takazoe (448) found that the vitamin requirements of the two strains he studied were the same but that amino acid requirements were slightly different. Both strains grew well in a medium containing salts, glucose, purines and pyrimidines, twelve vitamins and growth factors and thirteen amino acids.

**Table 9-2.** Comparison of biochemical tests of *B. matruchotii* grown aerobically and anaerobically[a].

| | Atmospheric Conditions | | |
| --- | --- | --- | --- |
| | Aerobic Stationary | Shaking | Anaerobic Stationary |
| Catalase | | | |
|   Aerobically derived | 195/195[b] | — — — | 22/60; 32/60± |
|   Anaerobically derived | 29/38 | | 6/38; 23/38± |
| Voges-Proskauer | 171/172 | 33/37 | 5/35[c] |
| Starch Hydrolysis | 67/67 | — — — | 17/71 |
| Maltose | 234/239 | 0/21 | 29/42 |
| Mannose | 122/123 | — — — | 27/42 |
| Raffinose | 78/159 | — — — | 17/56 |
| Salicin | 118/129 | | 0/42 |

[a] Data from Gilmour, M. N. (154).

[b] No. tests positive/No. strains tested.

[c] All thirty-five strains + when grown aerobically.

## D. SEROLOGY

Using FA procedures, we found that all strains of *B. matruchotii* tested stained with antiserum prepared against the type strain ATCC 14266 and there were no cross-reactions with *Actinomyces* or *Rothia*. Holmberg and Forsum (206) obtained similar results. A number of procedures including agglutination, precipitin and complement fixation tests have been used by various authors (251, 410, 415, 435) with the rather uniform conclusion that the antigenic composition of *Bacterionema* is complex but that the heterogeneity is not sufficient to establish serological groups or types. These various studies reported that no cross-reactions were demonstrable between *B. matruchotii* and species of *Actinomyces, Corynebacterium, Leptotrichia, Mycobacterium* or *Nocardia*.

# 10

# Ecology

With one exception, the *Actinomycetaceae* are associated with man and animals. *Actinomyces humiferus* is the only species which has been isolated exclusively from the soil, and it is the numerically predominant inhabitant of certain soils with a relatively high organic content. Numerous attempts to isolate other species of *Actinomyces* from soil have met with failure. Therefore, the unusual habitat of *A. humiferus* is one reason for questioning its classification as an *Actinomyces*.

The other species of *Actinomyces, Arachnia, Rothia* and *Bacterionema* are resident in the oral cavity of man or animals. Bergey (24) was the first to find that filamentous bacteria could adhere to the teeth and were involved in the formation of plaque. Then, Lord (290) demonstrated that oral actinomycetes were pathogenic for animals. Since then, numerous authors have demonstrated *Actinomyces* (33, 110, 206, 213, 425), *Arachnia* (79, 206, 425), *Rothia* (206, 386, 425) and *Bacterionema* (93, 157) in the human oral cavity. *Nocardia* has not been verified as a normal inhabitant of the mouth (organisms reported as *Nocardia* are usually *Rothia*). None of these genera have been reported as normal inhabitants of the intestinal tract. Occasional isolation of *A. israelii* from feces has been reported (447), but this probably represents its occurrence as a transient rather than a permanent inhabitant.

Considered individually, there are some differences in the occurrence of the *Actinomyces* species. *A. bovis* was first described (37) in material from the jaw of a cow and has been primarily associated with lumpy jaw in cattle. The source of the infection in cattle has not been established, but it is assumed that it is of endogenous origin. It should be emphasized that *A. bovis* has not been isolated from humans so that it seems to be essentially an animal parasite. *A. israelii, A. naeslundii, A. viscosus* and *A. odontolyticus* are all inhabitants of the oral cavity of man, and all except *A. odontolyticus* have been shown to cause infections in man. *A. odontolyticus* and *A. naeslundii* have not been reported from animals, but animal infections with the other two species are known. *A. israelii* has been found in infections in cattle (136) and swine (128, 173) and *A. viscosus* in infections in dogs, goats and swine (140). *A. viscosus* has also been isolated from spontaneous periodontal infections in hamsters (209) and from cervical plaque of rats (231). The provisional species, *A. suis*, has so far only been found in infections of animals (128, 173). The failure to find this relatively unknown species in man may be due to lack of recognition so it is impossible to state at present that it does not occur in the human oral cavity. The extent to which *Actinomyces* species occur as a part of the normal flora in animals is not known as there have been no systematic surveys of their occurrence in either domestic or wild animals. The finding of natural infections with these organisms, coupled with what is known of the epidemiology of actinomycosis in man, suggests that *Actinomyces* are a part of the normal oral flora in a variety of animals.

*Arachnia propionica* was first isolated from human lacrimal canaliculitis (367) and causes typical actinomycosis in man (50). It was not found to be part of the normal flora of the lacrimal duct or the conjunctiva (372) but is a part of the normal oral flora in man (206, 425). It has not been isolated from animals.

*Rothia dentocariosa* was first isolated from carious dentine (344) and has been repeatedly isolated from the oral cavity (95, 206, 393, 425). It has also been isolated from blood, spinal fluid and abscesses, but its etiological relationship to human disease has not been established (34, 52).

*Bacterionema matruchotii* has been isolated from human saliva, plaque, gingival crevice and periodontal pockets (13, 158, 206, 216, 217). It has also been isolated from monkeys (73) but not from other animals. *B. matruchotii* has not been reported from clinical specimens or as a cause of disease in man.

*Bifidobacterium* species are present in the oral cavity, intestinal tract and vagina of humans (171, 374, 388). They have also been isolated from the intestine of rats, guinea pigs, calves, pigs and turkeys as well as from bees and the bovine rumen (388). *A. eriksonii (B. eriksonii)* has been established as a human pathogen (144). It is found in tonsils (33) and probably in the intestinal tract (as Dehnert's group IV bifidobacteria).

Evidence for the occurrence of *Actinomyces* species other than *A. bovis* in the human oral cavity is extensive. These studies have been complicated by the many difficulties involved in culturing and identifying the bacteria so that quantitative data are probably not very accurate; however, the presence and relative proportions of the bacteria have been adequately documented. It has been reported that gram-positive facultative and anaerobic bacteria (including *Actinomyces* and others) constitute 15-20% of the cultivable bacteria in saliva and on the tongue, 40% of those in plaque and 35% of those in the gingival crevice (439). Organisms specifically identified as *Actinomyces* averaged 14% of the cultivable flora in plaque (289) in one study. All studies show that there is a wide subject-to-subject variation in the occurrence of these bacteria.

The fluorescent antibody technique has been applied to the direct identification of *Actinomyces* in tonsils and in dental plaque and calculus. Blank and Georg (33) identified *A. israelii, A. naeslundii* or *A. eriksonii* in 27% of 116 tonsil specimens and confirmed the FA identification by cultures. In our laboratory, we have examined both plaque and calculus by direct FA and culture for species of *Actinomyces, Arachnia, Rothia* and *Bacterionema* (79, 425). The results are

Table 10-1. Occurrence of *Actinomycetaceae* in forty-five samples of dental calculus[a].

| Species | FA Positive | | Culture Positive | |
|---|---|---|---|---|
| | No. + | %+ | No. + | %+ |
| *Actinomyces*[b] | | | | |
| A. israelii | 43 | 96 | 23 | 51 |
| A. naeslundii | 36 | 80 | 34 | 76 |
| A. viscosus | 35 | 78 | 26 | 58 |
| A. odontolyticus | 21 | 47 | 17 | 38 |
| *Arachnia propionica* | 23 | 51 | 10 | 22 |
| *Rothia dentocariosa* | 28 | 62 | 28 | 62 |
| *Bacterionema matruchotii* | 36 | 80 | 29 | 64 |

[a] Data from Slack et al. (425) and Collins et al. (79).

[b] All samples negative for *A. bovis.*

shown in Table 10-1. At least two *Actinomyces* species were found in all specimens and at least one species of each of the other genera were also present in all but one specimen. The number of species per sample ranged from three to seven. In a later study of plaque of known age, we found similar results. Plaque was removed from ten subjects, 3, 7 and 14 days after a complete prophylaxis. The distribution of individual species was somewhat erratic, but all subjects had each of the four species of *Actinomyces* in at least one plaque specimen. *Rothia* was present in all samples from all subjects. *Arachnia* was found less frequently but occurred in at least one sample from each subject. *B. matruchotii* was present in nine of the ten subjects at some time, but one subject never had this species. Holmberg and Forsum (206) also identified these organisms in dental plaque using FA and culture. They found *R. dentocariosa, B. matruchotii, A. viscosus* and *A. naeslundii* in supragingival plaque while *A. naeslundii, A. israelii, A. propionica* and *P. acnes* were found in subgingival plaque.

These studies and many others in the literature have adequately demonstrated that species of *Actinomycetaceae* (except *A. bovis*) are a constant part of the plaque microflora. Our isolation studies and those of others (289, 437) also indicate that not all the isolates from plaque which resemble *Actinomyces* can be satisfactorily identified with one of the known species. Further study of these isolates is necessary to give an accurate picture of the types of gram-positive facultative rods present in the oral cavity.

# 11

# Actinomycosis

## A. ACTINOMYCOSIS IN HUMANS

**Introduction.** Israel (220) in 1878 introduced the term "actinomycosis" in his accurate description of a cervicofacial and thoracic case of the disease, although he relates that Von Langenbeck in 1845 described the first case. Additional clinical descriptions soon followed along with the isolation of *A. israelii* by Bujwid (63) and a good description of the organism by Wolff and Israel (479).

As presently recognized, actinomycosis (82, 107, 111, 178, 442) is a chronic granulomatous disease characterized by suppuration, abscess formation and draining sinuses. Clinically, the disease is frequently designated as cervicofacial, thoracic and abdominal. For convenience, these designations will be used in this chapter but expanded to include infections of the central nervous system, bone and skin.

**Cervicofacial (118, 136, 188, 189, 197, 219, 236, 335).** This is a subacute or chronic granulomatous disease (Figures 11-1, 11-2, 11-3) in which the onset may go unnoticed until a persistent swelling develops (usually in the parotid or mandibular region). Then one or more draining sinuses develop, discharging a yellowish, thick to serous exudate which if carefully examined usually contains the characteristic sulfur granules. These granules should be examined microscopically for the presence of gram-positive filaments and hyaline clubs, although the latter are not always present (197, 473). The overlying skin often has a reddish-purple cast. There is rarely involvement of the regional lymphatics, but the infection may spread by direct continuity or via the bloodstream (175, 189). A periostitis may develop but direct invasion of the bone is rare, although it has been reported for the maxilla, mandibles and cranium (162, 335). Areas that have been reported as being primarily involved include: cheek, neck, parotideomasseteric, submandibular, retromandibular, mastoid bone, sinuses, parotid gland, thyroid, tongue, lips and ears (45, 66, 99, 123, 161, 262, 276, 302, 403, 443, 462). There is frequently a history of dental decay, tooth extraction, fracture of the jaw or other types of trauma. Basically, any circumstance which breaks the normal epithelial surface may provide devitalized tissue and anaerobic conditions for the growth of the *Actinomyces*. In brief, the clinical characteristics are: 1) pain (some are painless), 2) trismus of the masseters, 3) firm swelling, 4) fistulas, and 5) yellowish granules.

**Thoracic (80, 126, 376, 428, 459).** This form may result from: 1) inhalation or aspiration of the *Actinomyces* present in the oral cavity (in plaque, calculus or tonsillar crypts) via the tracheobronchial tree; 2) organisms being carried into the lung by a foreign body; 3) hematogenous extension; or 4) extension of an existing infection, such as cervicofacial actinomycosis,

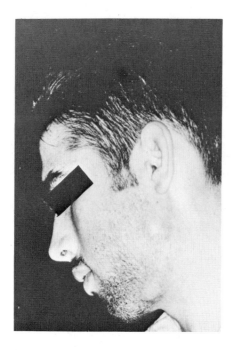

**Figure 11-1.** Early cervicofacial actinomycosis.

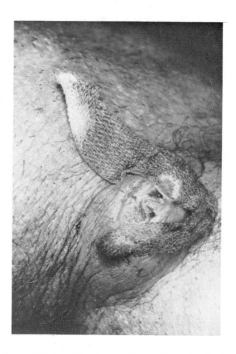

**Figure 11-2.** Draining sinus of cervicofacial
actinomycosis.

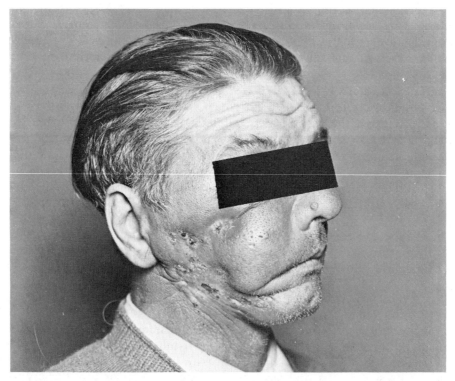

**Figure 11-3.** Cervicofacial actinomycosis. (Courtesy of L. Weed, Mayo Clinic,
Rochester, Minnesota.)

down tissue planes of the neck to the mediastinum or abdominal actinomycosis perforating the diaphragm. The primary lesion, which is seldom seen, may be in the bronchioles, peribronchial tissue or in the lung parenchyma. The disease progresses by direct continuity from lobe to lobe forming multiple abscesses with intervening dense fibrous tissue. There may be direct or indirect extension to the heart (105, 487). The pleura may be involved, becoming thickened and fibrosed; however, empyema is not common. The infection may spread through the chest wall producing multiple sinuses which tend to heal and then reopen discharging more pus (Figure 11-4). Sulfur granules may be observed in such exudate as well as in the sputum. The disease is sometimes divided into broncho, pleuro or pneumoactinomycosis with the principal symptoms being chest pains, fever, cough with expectoration and weight loss. Hemoptysis is not common (15). Clinically, actinomycosis may resemble tuberculosis, pulmonary abscess or bronchiogenic carcinoma (324, 378). With or without microscopic or cultural evidence of an actinomycete, it is often necessary to do an exploratory thoracotomy to provide a differential diagnosis and to rule out an underlying tumor. The x-ray changes are not characteristic, but a pulmonary infiltrate with bone

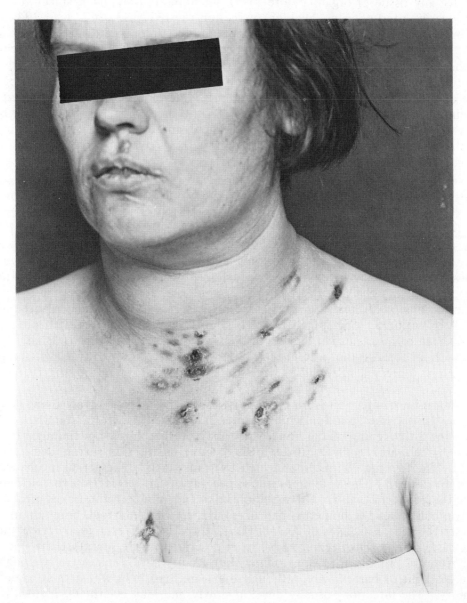

Figure 11-4.    Thoracic actinomycosis. (Courtesy of L. Weed, Mayo Clinic, Rochester, Minnesota.)

changes in ribs (proliferation or destruction) and lesions that extend across anatomical barriers should be highly suggestive of actinomycosis (125, 411) or a tumor. It is possible that these organisms are involved in pulmonary infections more frequently than ordinarily considered as Kay (237) isolated *Actinomyces* from 104 of a total of 240 patients with bronchopulmonary infections, but the etiological role the *Actinomyces* actually played was difficult to assess as other organisms such as spirochetes, fusiform bacilli and streptococci were usually present. Nevertheless, *Actinomyces* should be considered as a possible etiological agent in any chronic lung disease (180).

**Abdominal (3, 103, 120, 357, 380, 400).** This type of infection usually develops as a result of an acute perforative gastrointestinal disease or after trauma (surgical or accidental). The gastrointestinal disease could include any one of a variety of ulcerative diseases or appendicitis. Putman et al. (380) reported that 72% of 122 cases of abdominal actinomycoses were preceded by an attack of acute appendicitis, usually perforative. Because of this latter involvement, actinomycotic infections develop more frequently in the right lower quadrant of the abdominal wall. Extension from such foci is usually by direct continuity intraabdominally (could be hematogenous) to involve any tissue or organ (Figure 11-5), including muscle, liver, spleen, kidney, fallopian tubes, ovaries, uterus, testes, bladder, or rectum (8, 131, 133, 167, 176, 294, 349, 398, 441, 483). In many instances, the exact route or source (actinomycetes have not been reported as part of the normal flora of the intestinal tract) of the infection is difficult to trace. Direct invasion of the intestinal or stomach wall is only rarely observed (463, 477). Likewise, bone is not commonly involved, although vertebral invasion has been recorded (116). The patient becomes aware of such a complication usually through the development of an indurated mass (may simulate carcinoma) at the site of the operation or in the flank or pelvis which then begins to soften and develop one or more draining sinuses. These sinuses are sometimes the first reported symptom. The systemic symptoms are initially mild with a slow but progressive development of fever, malaise, weakness and pain.

**Central Nervous System.** Fetter et al. (122) indicates that there are over 100 reports of actinomycosis of the brain and spinal cord, but of these only twenty were verified by cultures. Involvement of the CNS usually results from direct extension of cervicofacial actinomycosis or from a hematogenous spread of a thoracic or abdominal locus. Either of the latter two may extend directly into the spinal canal (38, 379). In certain instances, the spinal column is directly involved (83, 485). The predominant symptoms are usually directed toward the primary site of infection, but the patient may complain of headache, show evidence of increased intracranial pressure or symptoms of focal seizures, hemiparesis, aphasia, ataxia or abnormal reflexes. The organisms may be observed directly in smears from the spinal fluid and may also be identified directly by FA or cultured. The spinal fluid cell count is usually elevated with predominantly polymorphonuclear or mononuclear leucocytes. In addition, the pressure is usually normal but may be high with an abscess; the protein may be normal but is frequently elevated; and sugar is usually moderately depressed. Patients may respond to antibiotics and/or surgery, but mortalities have been high.

**Bone.** Primary bone infection is rare. In most cases, the osseous involvement is an extension from an adjacent soft tissue infection which results in a periostitis with a stimulation of new bone formation causing a thickening of the bone which may be the only x-ray finding. If the bone is actively invaded, there are usually localized areas of bone destruction surrounded by increased bone densities. The foci may be interconnected by sinus tracts. The actual mechanism of bone destruction is not known. The x-ray appearance can vary from rarefaction to thickening and sclerosis resembling a bone tumor. The mandible, ribs and spine are the bones most frequently involved, but infections of the humerus, phalanx, knee, maxilla, mastoid, cranium and others have been described (18, 116, 248, 302, 310, 335). Cope (83) gives a good review indicating that sixty-six cases of bone involvement have been described in the literature through 1951.

**Skin and Wound.** Primary infections of the skin usually have a history of trauma resulting from a human bite, fistfight or barbed wire, although there are reports in which there is no obtainable history of injury. The infections usually develop slowly and may eventually involve the subcutaneous tissue, muscle and even bone with diagnosis dependent upon the presence of sulfur granules, filaments in gram stains or positive cultures. Treatment is usually successful with surgi-

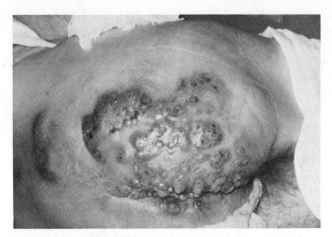

**Figure 11-5.** Ano-rectal actinomycosis. (Courtesy of W. A. Welton, WVU Medical Center.)

cal drainage and/or antibiotics. One of the earliest case descriptions is that of Cope (81) as indicated by Montgomery and Welton (323) in their case reviews. Cullen et al. (88) describe five cases in military personnel following mortar or bullet wounds and one after an injection, but *Nocardia* seems to have been isolated from three of these cases. Leavell et al. (269), using photographs and biopsy material, indicate that cutaneous actinomycosis goes through the sequential development of abscess formation, granulation, sinus tract formation and scarring.

**Etiology.** *Actinomyces israelii* is by far the most common species involved causing the classical actinomycosis with sulfur granules, but *A. naeslundii, A. viscosus* and *A. eriksonii* as well as *Arachnia propionica* may cause clinically similar infections. *A. bovis* has not been isolated from human infections. Infections without sulfur granules have been reported and may even be more frequent than the classical actinomycosis. *A. viscosus, A. naeslundii* and *A. israelii* may also be directly or indirectly involved in the development of caries and periodontal disease (437, 439).

Holm (204, 205) introduced the concept that human actinomycosis was a multiple infection arising through synergism between anaerobic actinomycetes and other organisms including *Actinobacillus actinomycetemcomitans, Bacteroides,* anaerobic streptococci and others. This concept has been supported by Lentze (279) and Hertz (197). There is little question that other organisms can be introduced into the tissue along with *Actinomyces* and thus add to the pathological process. However, there is no doubt that when one has actinomycosis an *Actinomyces* is present. Also, the disease with sulfur granules can be produced in experimental animals (77, 136). Thus, these organisms are the etiological agents of actinomycosis.

**Source.** The oral cavity is the principal habitat of these organisms. This was suggested by Israel (221, 222) particularly after he noted a piece of a tooth embedded in a pulmonary lesion. Bergey (24) in 1907 demonstrated filamentous organisms on the tooth surface and Lord (290, 291) injected material from teeth and tonsils intraperitoneally into guinea pigs producing abscesses even with sulfur granules. Emmons (110) in 1938 isolated anaerobic actinomycetes from tonsils, and this was later confirmed by Slack (416), Blank and Georg (33) and most recently by Hotchi and Schwarz (208) who specifically identified *A. israelii* and *A. naeslundii* in tonsillar sections. Recently in our laboratory (425) using the fluorescent antibody technique, we demonstrated the consistent presence of *A. israelii, A. naeslundii, A. odontolyticus* and *A. viscosus* in dental plaque and calculus. *A. bovis* has not yet been isolated from the human oral cavity.

With these organisms in the oral cavity, it is understandable how they can gain entrance into the tissues of the face and neck or be carried into the lungs via the tracheobronchial tree. However, as these organisms have not yet been reported among the normal inhabitants of the intestinal tract, although they must be transient occupants, it is more difficult to explain the source of infection in abdominal actinomycosis. It seems most likely, as transient occupants, they gain entrance through trauma, ulcerative or perforative gastrointestinal disease or are carried by the bloodstream to an injured or thrombosed site.

In relation to source of infection, it should be emphasized that the disease is *not* contracted from animals and that farmers or agriculture workers do *not* contract the disease more frequently than urban dwellers.

**Histopathology.** The basic microscopic picture in actinomycosis is suppurative but varies from that of an acute abscess with polymorphonuclear leucocytes, lymphocytes, plasma cells, macrophages (often foamy) and an occasional giant cell to that of a more chronic lesion in which proliferating connective tissue is the most conspicuous feature. Embedded in the purulent areas of the lesion will be the sulfur granules (may be found only after cutting multiple sections) with the central mass of intertwined filaments of the organism and the peripheral radiating clubs with their central filament of the actinomycete (see the discussion below). In some instances, the clubs are not present (473). Hematoxylin and eosin are the stains usually used for tissue sections, although a modification of the Gram's method (Brown and Brenn or MacCallum-Goodpasteur) may be necessary to clearly distinguish the filaments (111).

**Sulfur Granules.** The study of sulfur granules began in 1877 when Harz (190) observed the raylike structures in pus from the jaw of a cow. Observed microscopically and unstained, the granules gave the appearance of rays projecting out from a central mass of filaments and served as the basis for the name *Actinomyces* or "ray fungus." Since then, sulfur granules have been the hallmark of actinomycotic infections in animals and man.

When you take pus from a human or bovine infection and spread it out in a petri dish, the granules are often large enough to be seen with the naked eye but are more easily seen with a dissecting scope or a microscope at 100X. They have a yellowish appearance (hence the name "sulfur granule"), but there is no evidence that they contain sulfur. At this magnification, they appear irregular in shape and quite compact. For detailed examination, the next step is to transfer a granule, using a Pasteur pipette or inoculating loop, to a glass slide and gently crush it under a cover glass. Under low power the irregularity of the edge is distinct, and with reduced light there is a definite ray appearance at the periphery. Increasing the magnification to high dry or even to oil makes the "clubs" with their refractile or hyaline appearance become quite distinct, and one gains the impression that there is a central mass of filaments although they are difficult to distinguish (Figure 11-6). To further confirm the presence of the filaments, slide the coverslip off, heat fix, gram stain and observe the gram-positive branching filaments and diphtheroid cells. The organism can be specifically identified by doing direct FA staining on additional smears.

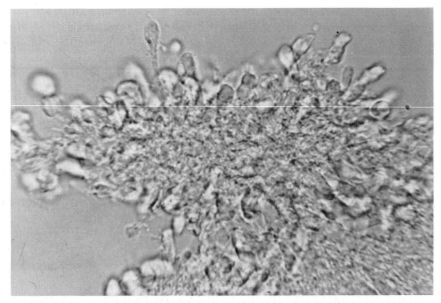

Figure 11-6.   Wet mount of sulfur granule from experimental *A. israelii* infection in a mouse. (Courtesy of L. Georg, CDC, Atlanta, Georgia.)

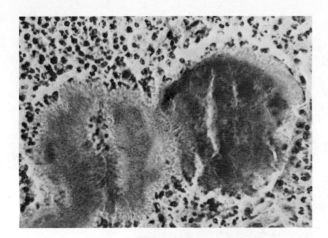
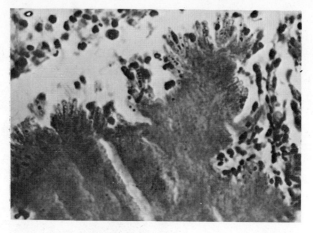

Figure 11-7.    Sulfur granule, human case.  Tissue section,    Figure 11-8.    The same sulfur granule showing peripheral
H and E, X450.                                                                            clubs.  Tissue section, H and E, X950.

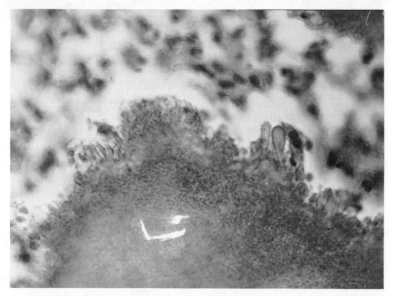

Figure 11-9.    Sulfur granule, human case, tissue section, H and E, X950.
(Courtesy of L. Weed, Mayo Clinic, Rochester, Minnesota.)

If tissue containing sulfur granules is fixed, sectioned, stained with hematoxylin-eosin and examined under oil immersion, one observes at the periphery the eosinophilic clubs (Figures 11-7, 11-8, 11-9).  With careful focusing, a basically stained filament will be observed approximately in the center of each club.  At the base of the clubs and occupying the central portion of the granule, one will see the intertwined, basically stained filaments as well as diphtheroid cells.  The filaments (Figure 11-10) are more easily seen if the granules are stained by the Brown-Brenn or MacCallum-Goodpasteur modifications of the Gram stain (111).  Such formaldehyde-fixed and paraffin-embedded sections can be stained directly by FA and the organisms specifically identified (208).  The organisms in the granules are non-acid-fast by the Kinyoun method, but variable results are obtainable with the Putt modification, and this should *not* be used if there is a question of differential diagnosis between *Actinomyces* and *Nocardia*.  There are conflicting reports on whether or not *Actinomyces* are PAS positive.  Hotchi and Schwarz (208) have a good discussion on the staining characteristic of these granules.

It is to be emphasized that the granules directly observed in pus should not be diagnosed as *Actinomyces* until the morphology is determined by staining and the species is specifically identified by FA or by culture, the reason being that other organisms or even masses of pus cells can

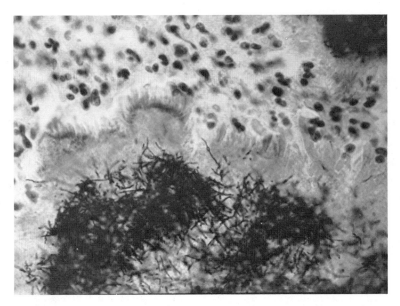

Figure 11-10. Sulfur granule, human case. Tissue section, gram's stain, X950. Note the stained filaments. (Courtesy of L. Weed, Mayo Clinic, Rochester, Minnesota.)

give a similar appearance at 100X or even under oil (473). Bacteria that can produce granules with clubs during the course of an infection in man are listed in Table 11-1.

In the literature there are conflicting statements as to whether or not *N. asteroides* can produce granules with hyaline radiating clubs, but according to Emmons et al. (111) they have not been observed. In our laboratory, we have not observed them in experimental infections in mice. They have been reported for *N. caviae*. It is true, however, that sometimes one sees a mass of

Table 11-1. Bacteria producing granules with clubs in human tissue[a].

| Organism | Disease | Gram | Acid-fast | Aerobic | Anaerobic | References |
|----------|---------|------|-----------|---------|-----------|------------|
| *Actinomyces israelii*[b] | Actino-mycosis | + | − | − | + | 208 |
| *Nocardia brasiliensis* | Mycetoma | + | + | + | − | 297 |
| *Streptomyces madurae*[c] | Mycetoma | + | − | + | − | 111 |
| *Staphylococcus aureus* | Botryo-mycosis | + | − | + | − | 440 |

[a] *A. bovis* and *Actinobacillus lignieresi* produce such granules in cattle. *A. lignieresi* only rarely involves man.

[b] Granules but without clubs have been reported for other species of *Actinomyces*, but more careful investigation might well demonstrate clubs.

[c] Other species include *S. pelletieri, S. somaliensis* and *S. paraguayensis.*

intertwined filaments without clubs, and differentiation is dependent upon the demonstration of acid-fastness, a negative FA with *Actinomyces* conjugated antiserum or isolation and specific identification of the organism. This differentiation becomes very important in establishing a diagnosis in man because sulfonamides and not antibiotics are the treatment of choice for *Nocardia asteroides* infections (175).

There have been numerous reports on attempts to stimulate *in vitro* club formation, but the positive report by Bayne-Jones (19) in 1925 has not been repeated. Likewise, there have been numerous proposed explanations for the *in vivo* formation of clubs. It has been suggested that the clubs consist of deposits of host-derived proteins on the surface of the filaments. In contrast, it has been proposed that the clubs are protein or polysaccharides synthesized by the organism which accumulate on the surface of the filaments. Crawford (87), with experimental evidence, indicates that clubs consist of polysaccharide produced by the organism combined with cationic antimicrobial polypeptides from the host's degenerating granulocytes. Regardless of the mechanism of formation, hyaline clubs are only rarely seen when the organisms grow free, as for example in the tonsillar crypts, but are demonstrable when tissue is invaded in man, cattle or experimental animals. Thus, it seems logical to assume that they are formed in response to a host-parasite interaction and probably function as a mechanism of resistance to phagocytosis or other immunological defense mechanisms of the host.

Recent studies of sulfur granules elucidating their chemical composition began when Widra (476) did histochemical studies on granules from a human lung and showed that the clubs appeared to be a highly polymerized, basic protein rich in guanidyl, indole and phenolic groups but lacking protein-bound, sulfhydryl groups. Pine and Overman (371) analyzed bovine sulfur granules and concluded that the clubs represented a polysaccharide-protein complex excreted by the organism and containing a high concentration of calcium phosphate. These observations were extended by Frazier and Fowler (129) using x-ray defraction procedures which showed that the granules were a poorly crystalized apatite of which the major constituents were CaO and $P_2O_5$ along with oxides of Na, K, and Mg. As indicated above, Crawford (87) found the cationic polypeptides liberated by the host's degenerating granulocytes. Thus, from the observations reported to date, it can be concluded that the hyaline clubs are a complex material of a polysaccharide-protein nature containing a variety of salts, but a more extensive and detailed analysis is required to better elucidate the exact chemical nature of this material.

**Incidence.** The first cases of actinomycosis reported in the United States were by Murphy (328) in 1885, and a tabulation of cases was made by Ruhrah (397) in 1899 and by Erving (117), giving a total of 103 recorded cases through 1902. A most extensive survey was made by Sanford (399, 401) in which he even canvassed physicians and hospitals, collecting 575 cases giving a total of 678 cases in the U.S. through 1923. More recently, Kolouch and Peltier (250) collected 163 cases from the literature along with thirty-eight new cases providing a total of *879* cases reported in the literature through 1945. From this date to the present, it is difficult to sort out the reporting of duplicate cases so a meaningful tabulation of the total number of cases is not possible. From 1968-1972, each of the following references lists ten or more cases (50, 105, 107, 125, 126, 147, 180, 312), and in each of these years there are an additional number of reports on single or multiple cases.

Even though actinomycosis is not a reportable disease, anywhere from six to twenty-five cases have been reported each year (CDC Morbidity and Mortality Reports, Atlanta, Georgia) but more surprisingly the reported deaths have averaged twenty-five per year for the years of 1949-1969 (Table 11-2). The entire subject of epidemiology and incidence is reviewed in considerable detail by Slack (418) in 1974.

Actinomycosis is worldwide in distribution: for example, in England 902 mortalities were recorded from 1916-1935 (82) and 169 cases reported for the years 1966-1970 (428). In Australia, Durie (104) indicates the average annual rate in Sydney was ten from 1946-1956, and Roveda (395) observed sixteen cases during the years 1965-1971. In Denmark, Holm (204, 205) lists 785 patients, and Rud (396) reported on 234 cases during the years of 1945-1967. Lentze (278, 279) gives information about 1002 cases occurring in Germany. This topic is reviewed in considerable detail by Cope (82), Bonner and Bonner (39) and Slack (418).

**Table 11-2. Actinomycosis—Reported mortality from vital statistics of U.S. crude death rate, 1949-1969.**

| Year | Year Rank | Deaths | U.S. Resident Population[a] x 1000 | Reported Crude Death Rate Per 100 Million | Crude Death Rate Rank | d | $d^2$ |
|------|-----------|--------|------------------------------------|-------------------------------------------|-----------------------|-----|-------|
| 69 | 21 | 15 | 201,385 | 7.45 | 1 | 20 | 400 |
| 68 | 20 | 16 | 199,399 | 8.02 | 2 | 18 | 324 |
| 67 | 19 | 28 | 197,457 | 14.18 | 11 | 8 | 64 |
| 66 | 18 | 21 | 195,576 | 10.74 | 6 | 12 | 144 |
| 65 | 17 | 26 | 193,526 | 13.43 | 9 | 8 | 64 |
| 64 | 16 | 31 | 191,141 | 16.22 | 16 | 0 | 0 |
| 63 | 15 | 35 | 188,483 | 18.57 | 17 | −2 | 4 |
| 62 | 14 | 17 | 185,771 | 9.15 | 4 | 10 | 100 |
| 61 | 13 | 21 | 182,992 | 11.48 | 7 | 6 | 36 |
| 60 | 12 | 28 | 179,979 | 15.56 | 14 | −2 | 4 |
| 59 | 11 | 27 | 177,135 | 15.24 | 13 | −2 | 4 |
| 58 | 10 | 21 | 174,149 | 12.06 | 8 | 2 | 4 |
| 57 | 9 | 15 | 171,187 | 8.76 | 3 | 6 | 36 |
| 56 | 8 | 24 | 168,088 | 14.28 | 12 | −4 | 16 |
| 55 | 7 | 23 | 165,069 | 13.93 | 10 | −3 | 9 |
| 54 | 6 | 17 | 161,884 | 10.50 | 5 | 1 | 1 |
| 53 | 5 | 30 | 158,956 | 18.87 | 18 | −13 | 169 |
| 52 | 4 | 25 | 156,393 | 15.99 | 15 | −11 | 121 |
| 51 | 3 | 37 | 153,982 | 24.03 | 21 | −18 | 324 |
| 50 | 2 | 36 | 151,868 | 23.70 | 20 | −18 | 324 |
| 49 | 1 | 32 | 149,304 | 21.43 | 19 | −18 | 324 |
| | | | | | | | 2472 |

Correlation, Mortality vs. Year. Spearman's $r = 1 - \dfrac{6\Sigma d^2}{n(n^2-1)} = 1-6(2472) = 1-1.605 = .605$

[a] Statistical Abstract of the U.S. 1972, Table No. 2, p. 5, 93ed, U.S. Commerce Dept. Government Printing Office, Washington, D.C.

From these reports it can be concluded that "Actinomycosis occurs throughout the world and that it is neither a rare nor a common disease."

**Age, Sex and Case Distribution.** Actinomycosis has been reported in persons ranging in age from 28 days to 82 years, with the median falling between 20 and 30 years (399, 401). Most tabulations which include sex ratios indicate that the ratio is between two to four males per female, and Sanford (399, 401) states that 80% of his 575 cases were males. The anatomical distribution of cases as reported by Sanford (399, 401) and Cope (82) showed approximately 60% cervicofacial, 15% thoracic, 20% abdominal and 5% other types.

# TREATMENT

**Introduction.** As incision and drainage of pus is an ancient surgical remedy, it was no doubt applied to actinomycosis long before the disease was described (250), and in fact Israel (221) mentions rib resection and drainage in one of his cases. Waring (472) emphasized the principles of curreting, drainage and irrigation with an antiseptic solution. Then Wangensteen (470, 471) recommended the use of energetic surgery to ensure the removal of all devitalized tissue, and this continues to be of primary importance in any case of actinomycosis (180).

It is probably safe to say that almost any therapeutic agent that has been used to treat a bacterial or fungus infection has been tried in the treatment of actinomycosis. This included iodides, thymol, copper sulfate, hydrogen peroxide, silver nitrate, arsenicals, vaccines and irradiation (15, 25, 250, 331, 354, 380). Of these, iodides (250, 465) and irradiation (187, 429) had been the most extensively used until the advent of the sulfonamides.

**Sulfonamides.** Satisfactory treatment of actinomycosis became possible with the introduction of the sulfonamides. Poulton (375) in 1937 briefly mentioned the successful treatment of a case of abdominal actinomycosis, and the next year Walker (466) gave a more detailed account of the treatment of a similar case. Lyons et al. (293) reviews the early cases stressing the fact that a prolonged treatment is necessary to ensure recovery and indicates that sulfadiazine is the drug of choice. Then Kolouch et al. (250) lists fifty-three cases reported in the literature with a cure rate of about 65%, but he points out that in most instances the sulfonamides were used in conjunction with surgery. Sulfonamides continue to be used as supportive therapy (66).

**Antibiotics.** Soon after penicillin became available, Abraham et al. (1) in 1941 reported that one strain of *A. bovis* was sensitive to penicillin. Florey et al. (124) included a thoracic and abdominal case of actinomycosis in their series of fifteen patients treated with penicillin, but neither case responded to the treatment using about 400,000 units of penicillin. Then Herrell (195) listed four cases of actinomycosis with two cures treated with penicillin at the Mayo Clinic, and by 1945 some thirty-three treated cases had been reported in the literature (250). In 1948 Nichols and Herrell (340) reported on sixty cases with a high percentage of cures but pointed out that the failure rate was still 4-18%. The failure rate could in part be due to: 1) inadequate doses of penicillin, 2) inadequate blood supply at site of infection, 3) multiple abscesses surrounded by granulation tissue which interferes with penetration, 4) secondary invaders which interfere with the action of penicillin, or 5) the difficulty of penetrating granules with clubs. In 1949 Sanford and Barnes (400) discussed these problems and indicated that to reduce the failure rate massive doses of penicillin should be administered for a prolonged period of time. Harvey et al. (189) supported the combination of initial massive penicillin therapy, extensive surgical excision of infected tissue and continued penicillin for 12-18 months after excision. These concepts were reconfirmed by Peabody and Seabury (353, 354) and by Hartley and Schatten (188) in 1973.

As other antibiotics or chemotherapeutic agents were introduced they were used in treating various cases of actinomycosis usually because the patient had not responded to therapy or because the penicillin or sulfonamides could not be tolerated. These agents have included streptomycin, tetracyclines, chloramphenicol, erythromycin, ampicillin, cephalothin, isoniazid and stilbamidine; at the present time clindamycin and lincomycin are gaining in use (66, 85, 195, 267, 268, 287, 304, 313, 314, 319, 389, 481). Lerner (280) tested seventy-four strains of actinomycetes in semisolid antibiotic agar and reported that erythromycin, cephaloridine, clindamycin, minocycline, penicillin G and rifampin were very active *in vitro*. Amphotericin B, nystatin and griseofulvin are ineffective in the treatment of actinomycosis. In addition to these, surgery combined with hyperbaric oxygen has been successfully used in a case of anorectal actinomycosis (303).

**Vaccine Therapy.** The earliest report on the use of vaccines is that of Wynn (482) in 1908, although there is uncertainty about the identity of the organism he used. There are then a number of early isolated reports on the use of vaccines, but none as extensive as Colebrook (76), who used suspensions of dried organisms in twenty-three patients with success in a number of cases of cervicofacial actinomycosis. Neuber (339) in 1940 used a phenolized suspension of the organisms. Negroni (337) prepared a broth filtrate following the method of tuberculin production, giving increased concentrations at 2-day intervals. The most extensive and presently continued use of vaccines is in Germany with the use of the heterovaccine developed by Lentze (277, 279) at the Institute for Hygiene in Cologne. This vaccine contains killed suspensions of *A. israelii*, *A. propionica*, *Actinobacillus actinomycetemcomitans*, *Bacteroides melaninogenicus* and perhaps others. It is used as an initial excitant to promote drainage and then in increasing concentrations at 5-day intervals for over 80 days. Buchs (62) has reported a series of cases successfully treated with such a vaccine without the use of antibiotics.

**Lacrimal Canaliculitis and Conjunctivitis.** In 1875 Cohn (75) described a branching filamentous organism in lacrimal concretions and named the organism *Streptothrix foersteri.* It turned out that the generic name *Streptothrix* had been previously applied to another organism so it could not be used for the actinomycetes. Nevertheless, the name is still perpetuated in ophthalmological literature, and even today one sees reference to streptothrix infections or streptothricosis. However, such terms should not be used, and the organism should be termed *Actinomyces* and the disease called actinomycosis or lacrimal canaliculitis due to *Actinomyces,* providing that an actinomycete has been identified. The filamentous organisms that have been implicated include *A. israelii, A. odontolyticus, Arachnia propionica* and not too frequently *Nocardia asteroides* (109, 368, 404).

The *Actinomyces* and *Arachnia* present in the oral cavity are in all likelihood carried into the eye by the person's own fingers or from aerosols or possibly refluxed from the mouth into the lacrimal duct. In rare instances, the eye is infected as the result of direct extension from an existing infection (372). *Nocardia* are soil inhabitants so they could be carried on hands or by foreign bodies.

Such patients often recall having a conjunctivitis followed by the development of an intermittent creamy discharge from the corner of one of the eyes. This may have persisted for months or 2-3 years or more. When the canaliculi are probed, there will be found one to several yellowish concretions which can be removed by pressure or with a curette: in some instances these are adherent to the wall. The concretions will vary in consistency from soft to quite hard or grainy. These concretions can be smeared and gram stained to demonstrate the organisms, and they can also be cultured. In our laboratory, we have had concretions from two such cases re-

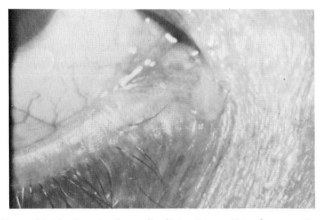

Figure 11-11. Lacrimal canaliculitis. Direct FA of concretions shown below.

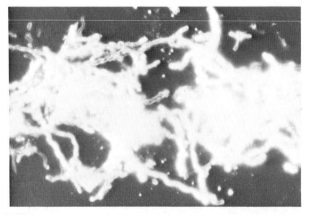 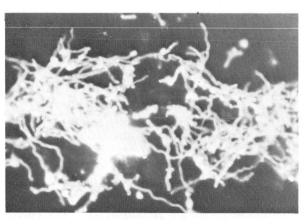

Figure 11-12. Direct smear of concretions from a case of lacrimal canaliculitis stained with FITC conjugated antiserum for *A. propionica.* Darkfield, X1200.

Figure 11-13. Same field as 11-11 with UV illumination showing many filaments stained with the antiserum, X1200. This specimen also contained *A. israelii.*

ferred to us and with the use of direct FA identified the organisms as *A. israelii* plus *Ar. propionica* (Figures 11-11, 11-12, 11-13) in one case and *Ar. propionica* plus *A. odontolyticus* in the other case. These identifications were made and a report forwarded to the physician while the patient was still in the clinic. These were then verified by cultural isolation of the organisms. Removal of the concretions followed by irrigation with antibiotics usually results in complete recovery. Pine et al. (368) reported thirty such cases from the literature.

## CARIES AND PERIODONTAL DISEASE

**Introduction.** The first recorded drawings of bacteria were made in 1863 by Antony van Leeuwenhoek from his microscopic examination of material from his teeth. Although he cleaned his teeth by rubbing with salt, he noted, "Yet notwithstanding, my teeth are not so cleaned thereby, but what there sticketh or growth between some of my front ones and my grinders, a little white matter, which is as thick as it 'twere batter." He collected this material, mixed it with rainwater and spittle, examined it with his crude microscope and indicated, "I then most always saw, with great wonder, that in the said matter there were many very little living animalcules, very prettily a-moving." He then examined tooth scrapings from a number of persons and noted that although the numbers of organisms varied they were similar in size and shape (98).

The occurrence of massive numbers of bacteria in various areas of the mouth, including saliva, the tongue and the gingival crevice as well as the tooth surface has since been confirmed by numerous investigators. As early as 1847, the suggestion was made by Ficinus that microorganisms caused dental caries. This was given further emphasis by Williams in 1897 when he demonstrated that microorganisms formed a "felt-like mass" adhering to the enamel and postulated that caries began where acid was formed by these masses of microorganisms. The term plaque for this deposit of microbes was apparently introduced by Black in 1898 (444).

**Plaque.** Starting with a thoroughly cleaned tooth, within a matter of minutes salivary glycoproteins are selectively adsorbed to the enamel to form an amorphous membranous layer or pellicle which eventually becomes a highly insoluble coating 0.1 to 10 $\mu$m thick. In a short time after its formation, bacterial microcolonies begin to develop on the pellicle. These come from bacteria present in the cracks and fissures or those adsorbed from saliva. The microcolonies increase in size and then coalesce to form a continuous bacterial layer. This dense, noncalcified bacterial mass firmly adherent to the tooth surface is referred to as dental *plaque* (Figure 11-14). Plaque contains a wide variety of bacteria including cocci, rods and filamentous forms which may reach a population of $10^{11}$ bacteria per gram of wet weight. There is evidence that the bacterial composition of the plaque varies from tooth to tooth in the same mouth and even in different sites on the same tooth. Plaque formation, both supragingival and subgingival, is the initial step in the development of caries and periodontal disease (153).

**11-14.** Massive plaque accumulation on hamster root surfaces. Animal infected with *A. viscosus*. (Courtesy of P. H. Keyes, NIDR, Bethesda.)

The bacteria which colonize the surface of the tooth require some type of adherence mechanism to allow them to remain in this location and not be washed out by salivary flow. Plaque bacteria must be able to attach directly to the tooth surface or to adhere to other bacteria already in the plaque. *Actinomyces* appear to possess both types of adherence mechanisms. *A. naeslundii* and *A. viscosus* synthesize extracellular polymers which may aid in their adherence to the tooth surface. These bacteria have been shown to aggregate in saliva and to adsorb to human enamel powder (152, 198). Interbacterial aggregation between *A. naeslundii* and streptococci (151) and the selective adherence of *Veillonella* to *A. viscosus* (153) have also been shown. Filamentous bacteria contribute markedly to the volume of plaque and present an enormous surface for the adherence of other bacteria (43). It has been shown in scanning electron microscopic studies of plaque that cocci adhere to filaments (Figure 11-15), giving a very interesting "corn-cob" appearance (153, 286).

**Caries.** After plaque is formed, the tooth surface beneath the plaque is subjected to a continuum of metabolic products from the bacteria. Of these products, the organic acids (lactic, acetic, propionic, succinic, formic) in lowering the pH to 5 or below initiate the process of demineralization of the enamel. There follows proteolytic degradation of the dentine resulting in a cavity which may extend and destroy most of the tooth. This is an oversimplification of the process as there continues to be controversy over the exact mechanism of demineralization and cavity formation. This controversy emphasizes the fact that these are complex processes and can probably occur in more than one way. Diet and heredity also play a role in caries susceptibility. Nevertheless, there is uniform agreement that caries do not form in the absence of plaque and that bacteria are involved.

Even after acid decalcification of the tooth enamel was recognized as a cause of caries, it was believed that all plaque microbes contributed to the process and that the bacterial etiology of

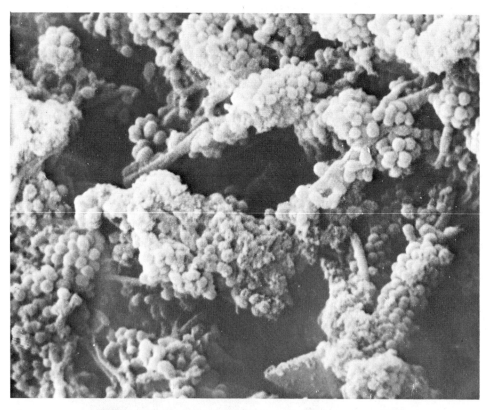

Figure 11-15. Scanning electron micrograph of human dental plaque showing "corn-cob" adherence of cocci to filaments. (Courtesy of Z. Skobe, Forsyth Dental Center, Boston.)

caries was nonspecific. In the past 10 years, evidence has accumulated implicating specific organisms as the causative agents of caries. Most of this information has come from the use of gnotobiotic or pathogen-free rats and hamsters. The evidence for the bacterial etiology of both caries and periodontal disease was reviewed by Gibbons in 1972 (148). Some evidence showing a relationship between the occurrence of certain cariogenic organisms and caries incidence in man (150) has been found. *Streptococcus mutans* has been firmly established as a causative agent of enamel caries (148, 408). Other streptococci and some lactobacilli may also initiate such lesions (392, 413). Filamentous bacteria, including *Actinomyces* and *Rothia,* have been implicated in root surface caries (Figure 11-16) in rats and hamsters (240) and have been isolated from human root surface caries (228). Filamentous bacteria have not usually been associated with enamel lesions (148), but Frank et al. (127) were able to produce fissure lesions in germ-free rats with both *A. naeslundii* and *A. viscosus.*

**Periodontal Disease.** The gingival crevice harbors literally billions of bacteria (1.3-2.1 x $10^{11}$ bacteria per gram wet weight and an estimated total of 16 mg wet weight in the overall gingival trough), but the intact gingival epithelium resists the potentially damaging metabolites of these microorganisms. However, when there is a localized increase in bacterial numbers as in excessive plaque accumulation, a shift in the kinds of bacteria or an injury (can be due to calculus accumulation), gingivitis occurs which can progress to periodontitis. The mechanisms that initiate this epithelial damage are not well defined and are probably multiple, but they could include: 1) bacterial enzymes (proteases, hyaluronidase, deoxyribonuclease, chondroitin sulfatase and collagenase), 2) endotoxins, 3) hypersensitivity reactions that initiate lymphocytic blast transformation and cytotoxicity, 4) bacterial competition for nutrients and 5) factors still to be recognized (437).

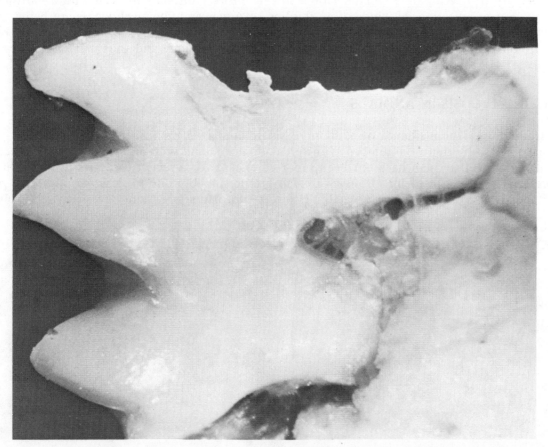

**Figure 11-16. Hamster infected with** *A. viscosus,* **mandibular first molar showing lesions involving only the cementum covered root surfaces. (Courtesy of H. V. Jordan, Forsyth Dental Center, Boston.)**

The gingivitis which is initially subclinical becomes clinical and even suppurative. The infection can then slowly spread to the deeper parts of the periodontium causing degeneration of collagen fibers, an apical migration of the gingival epithelium, formation of periodontal pockets which may develop into periodontal abscesses and eventually resorption of alveolar bone with the loss of teeth (65).

As plaque is a prime etiological factor in periodontal disease, one would expect the same type of bacteria to be present as in supragingival plaque. In the main this seems to be true, except that the environment is more anaerobic and greater numbers of *Bacteroides, Bifidobacterium, Clostridium, Veillonella, Treponema* and *Peptococcus* are found. Because the *Actinomyces* and most of the other filamentous bacteria are facultative, they will thrive under these conditions, adding their surfaces for adherence as well as their acids and enzymes to the biological pool having a deleterious effect on the gingival epithelium and the periodontum. Calcification of plaque leads to increased gingival irritation and inflammation. Filamentous bacteria contribute to the calcification, especially *B. matruchotii*, which can convert intracellular calcium to hydroxy-apatite (112).

The evidence that *Actinomyces* play a role in periodontal disease is: 1) they are universally present in gingival crevices, and 2) *A. viscosus* and *A. naeslundii* have been shown to initiate periodontitis (Figures 11-17 and 11-18) in hamsters and gnotobiotic rats (232, 438).

As the development of plaque with its bacterial population is responsible for both caries and periodontal disease (288), there is now a concerted effort to develop means of preventing plaque from developing or of erradicating plaque after its development. The most practical means of caries prevention at the present time is the use of sodium fluoride which binds with the hydroxy-apatite to make the enamel more resistant to decalcification. Diet control is another factor, and the possibility of developing a vaccine should not be entirely discounted even with a multiple etiology. For erradication, all kinds of antibiotics, enzymes and chemical agents are being tried. Chlorhexadine looks promising at the moment. In the near future, there will no doubt be developed biological and chemical methods which will control plaque and thus greatly reduce the incidence of both caries and periodontal disease.

## B. ACTINOMYCOSIS IN ANIMALS

**Introduction.** The incidence of actinomycosis in animals is greatest in cattle, although infections have been reported in a number of different domestic and wild animals. Natural infections in primates other than man have not been reported except for a recent case in a drill (7) due to *A. israelii.* The majority of infections in animals are presumably due to *A. bovis,* although *A. israelii* and *A. viscosus* have been isolated. However, unfortunately there have been relatively few

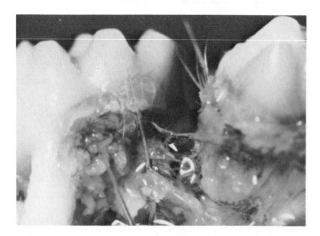

Figure 11-17. Periodontal lesions in rat monoinfected with a human isolate of *A. viscosus.* (Courtesy of P.H. Keyes, NIDR, Bethesda.)

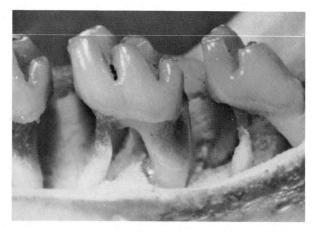

Figure 11-18. Periodontal lesions in a hamster infected with *A. naeslundii* N16. (Courtesy of P. H. Keyes, NIDR, Bethesda.)

good microbiological studies of isolates of actinomycetes from either normal or infected animals so our knowledge of these organisms is meager. This would be a fertile field for research.

**Bovine.** Bovine actinomycosis (lumpy jaw) is a chronic infectious disease (Figures 11-19 and 11-20) usually involving the mandible but sometimes the maxilla. It begins as a circumscribed, hard, immovable protuberance, usually in dorso-ventral alignment with the molar teeth. As the infection develops there is bone destruction and at the same time a stimulation of bone growth causing the proliferative osteitis. Sinus tracts then develop which discharge a yellowish purulent exudate. This exudate frequently contains firm, yellowish granules or "sulfur granules" which, if crushed and examined microscopically, will be shown to contain the branching filaments characteristic of the actinomycetes. The organism rarely invades the local lymph nodes, but there can be a gradual extension into the contiguous tissue or dissemination via the blood stream. The animal often remains relatively healthy unless there is impairment of mastication or breathing and then there is a marked loss of weight. Primary lung and soft tissue infection have been reported (29). For some reason, the incidence of lumpy jaw in the United States seems to be decreasing.

The source of infection is not established, but since *A. bovis* has not been isolated from the soil or plants it is no doubt endogenous. The organisms gain entrance into the tissue via a traumatic agent such as plant material (incidence increases when bearded grain straw or silage is fed) or a broken tooth. Direct transmission from animal to animal does not seem to occur, and even transmission from cow to calf has not been established.

Figure 11-19. Bovine actinomycosis (lumpy jaw). Involvement of the maxilla of a bull with some discharge. (Courtesy of M. L. Miner, Veterinary Science, Utah State University, Logan, Utah.)

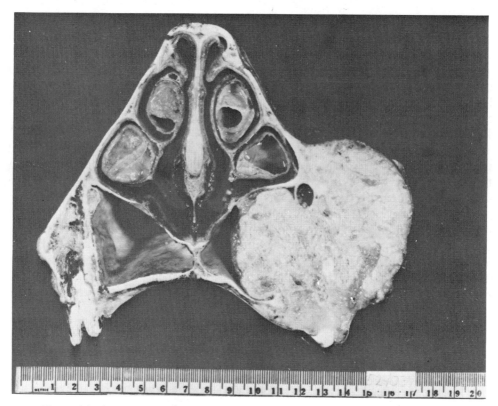

Figure 11-20. Bovine actinomycosis involving the paranasal sinuses. Transverse section through nasal turbinate and lesion. (Courtesy of C. R. Cole, Veterinary School, Ohio State University, Columbus, Ohio.)

*A. bovis* is the principal etiological agent, although infections with *A. israelii* have been reported (91, 243, 370). Such infections invariably harbor a mixed flora of microorganisms, and these organisms may increase the severity or promote extension of the infection; yet, without the actinomycetes, clinical disease does not develop. The treatment of bovine actinomycosis includes drainage and curettement with or without the use of iodides, penicillin, streptomycin or isoniazid (51).

Another infection of cattle which may resemble actinomycosis is *actinobacillosis* or "wooden tongue," caused by the gram-negative rod, *Actinobacillus lignieresii.* In this disease, the organism involves primarily the tongue, lymphatics and soft tissues of the neck. Bone is not invaded. Granulomas are formed in the tongue, which then becomes stiff and hard, giving the name "wooden tongue." Sinuses may form, and the yellowish exudate often contains "sulfur granules." Stained smears from these granules will show numerous gram-negative rods instead of the gram-positive filaments of the actinomycete granules. *A. lignieresii* is gram negative, nonmotile, no capsule, + catalase, + indole and nitrate, produces acid from glucose, maltose, mannitol and sucrose, no acid from xylose or inositol and is serologically distinct (454).

**Swine**. Magnusson (299) wrote an extensive article on actinomycosis in domestic animals and described at some length such infections in swine. The clinical picture seems to be variable with the udder, lungs or internal organs being more frequently involved but bone invasion has been described (454). Staphylococci or other organisms may cause similar infections. *A. israelii* (299), *A. viscosus* (140) and *A. suis* (128, 173) have been isolated from infections in swine.

Grässer (128, 173) cultured granules from udder actinomycosis of swine and isolated *A. israelii* from some animals and a new species which he designated as *A. suis* from other animals. A similar organism has been isolated by G. W. Robertstad (personal communication), and recently Franke (128) has further described such isolates. However, as Grässer's original cultures are not available, it cannot be determined whether all these isolates are the same.

**Dogs.** Canine actinomycosis has been described (140, 311), including infections of the soft tissue of the jaw and internal organs. Georg et al. (140) reports isolating *A. viscosus* from two different canine cases; however, other than this, there have been no studies of canine isolates so it is not known whether a particular species is involved.

**Other Animals.** Actinomycosis-like infections have been reported in sheep, goats, horses, cats, deer, moose, antelope and mountain sheep. Most of these are from clinical impressions or from histological examination of tissue, but again reports on isolation and identification of the isolates are lacking (96, 234).

## C. EXPERIMENTAL ACTINOMYCOSIS

*Actinomyces bovis.* Wright (480) in 1905 inoculated thirteen guinea pigs, two rabbits and one calf with cultures from two different cattle and demonstrated abscess formation with clubs in eleven of the guinea pigs and one of the rabbits. Then Magnusson (299) in 1928 inoculated thirty-two cattle subcutaneously with cultures or pus from cases of bovine actinomycosis and obtained localized infections with granules in eight of these animals. He also inoculated pigs, sheep, goats, dogs and horses but with essentially negative results. Meyer and Verges (317) in 1950 mixed cultures of *A. bovis* with mucin and injected mice intraperitoneally, obtaining abscesses in seventeen of twenty-one mice. Pine et al. (370) in 1960 produced abscesses in hamsters with a combination of intraperitoneal and subcutaneous injections but found that *A. bovis* was less pathogenic for hamsters than *A. israelii*. As with all attempts to produce infections with these organisms, the results are variable as to both strains of *Actinomyces* and animals, and, although localized and sometimes generalized abscesses can be produced, there rarely results a generalized fatal infection. A summary of selected reports on animal experimentation is given in Table 11-3.

*Actinomyces israelii.* Wolff and Israel (479) in 1891 were first to report experimental infections using cultures of human isolates. They injected rabbits, guinea pigs and sheep, obtaining demonstrable lesions with granules and filaments in all except the sheep. Naeslund (333) in 1929, using pure cultures from human cases, inoculated cattle, swine, rabbits and guinea pigs by a variety of routes. The guinea pig and swine experiments were negative, but lesions were produced in cattle and rabbits which gave positive cultures and some showed typical granules. Emmons (110) in 1938 had essentially negative results in guinea pigs using isolates from human tonsils, but Slack (416) in 1942 produced progressive infections in rabbits. However, Rosebury et al. (391) in 1944 had mostly negative results with rabbits. Meyer and Verges (317) in 1950 produced lesions in over 80% of a large series of mice following intraperitoneal injections with the organisms suspended in mucin. Geister and Meyer (135) in 1951 used the same technique with mucin and demonstrated the therapeutic effectiveness of penicillin and aureomycin in experimental infections in mice. Gale and Waldron (132) in 1955 modified this technique using mucin or broth injected at different sites and produced abscesses in 60% of the mice. Hazen et al. (192, 193) reported a high rate of infection with hamsters. Pine et al. (368) and Georg et al. (141, 143) infected both mice and hamsters, and, in an extensive report including histopathological studies, Brown and von Lichtenberg (53) infected 93% of young mice with rough strains of *A. israelii*. Infections have also been produced on the chorioallantoic membrane of chick embryos (355). A compiled list of animal experiments is given in Table 11-3.

Mathieson et al. (307) in 1935 raised the question of the role of hypersensitivity in experimental and human actinomycosis. They attempted to sensitize guinea pigs and rabbits but obtained either negative or minimal skin test reactions. They also did skin tests on five actinomycosis patients with negative results. Emmons (110) in 1938 also was unsuccessful in attempts to sensitize guinea pigs. In our own laboratory from time to time, we have tried a variety of antigens for sensitizing and skin testing both rabbits and guinea pigs, but the results have also been essentially negative. Thus, the evidence to date indicates that hypersensitivity does not play a major role in the development of actinomycosis. However, one should probably still be somewhat conservative and indicate that the question should remain open for a more extensive study with a wider variety of antigens and procedures.

**Table 11-3. Recent reports of experimental infections with *Actinomyces* and *Arachnia*[a].**

| References | Organism | No. Strains | Inoculum[b] | Mouse[c] | Hamster[c] | Abscesses with Filaments, Granules or + Cultures; Mortality Rare |
|---|---|---|---|---|---|---|
| 53, 56 141, 144 | *Actinomyces israelii* | 19 | Sal.Susp. .2-.5 ml IP | 9-30 g 1-240 days | | 181/189[d] |
| 135 317 | *Actinomyces israelii* | 14 | Sal.+Mucin 0.5 ml IP | 10-15 g 10-15 days | | 98/103 |
| 132 | *Actinomyces israelii* | 7 | Sal.+ Mucin .2 IP | Random 24 days | | 62/105 |
| 192 193 | *Actinomyces israelii* | 23 | Sal.Susp. .21-1 ml IP | | 500-800 g 3-21 days | 52/66 |
| 317 | *Actinomyces bovis* | 5 | Sal.+Mucin 0.5 ml IP | 10-15 g 15 days | | 17/21 |
| 56 | *Actinomyces bovis* | 1 | Sal.Susp. 0.5 ml IP | 5 wks. old 32 days | | 1/6 |
| 370 | *Actinomyces bovis* | 10 | Water Susp. 0.5 ml IP | | 4-6 wks. 10-30 days | 6/10 |
| 56, 78 141 | *Actinomyces naeslundii* | 13 | Sal.Susp. .2-5 ml IP | 15-30 g 10-32 days | | 111/129 |
| 141 144 | *Actinomyces eriksonii* | 5 | Sal.Susp. 0.5 ml IP | 15-30 g 10-28 days | | 34/49 |
| 141 | *Actinomyces odontolyticus* | 6 | Sal.Susp. 0.5 ml IP | 15-30 g 10-15 days | | 6/68 |
| 141 | *Actinomyces viscosus* | 4 | Sal.Susp. 0.5 ml IP | 15-30 g 10-15 days | | 23/24 |
| 56 141 | *Arachnia propionica* | 4 | Sal.Susp. 0.5 ml IP | 15-30 g 10-33 days | | 28/28 |

[a] This is a selective list in which the identity of the culture was established and in which a significant number of animals were used. A number of other reports indicate negative or minimal results particularly with rabbits and guinea pigs.

[b] This gives the suspending fluid used for the injection, the amount injected and the route. Consult the reference for the method of growing the organism. Also some authors included other routes of inoculation.

[c] The weight or age of the animal is given and then beneath this is the period of observation prior to sacrificing the animal.

[d] The number of animals showing abscesses/number of animals injected.

The quoted authors have amply demonstrated that pure cultures of *A. israelii* will produce experimental actinomycosis in which localized or progressive lesions develop. Typical sulfur granules with hyaline clubs are demonstrable. Smears consistently show gram-positive filaments which can be demonstrated and identified in direct smears with FA, and in most instances cultures are positive (Figure 11-6). However, pathogenicity for experimental animals is obviously strain variable, and along with this animals vary in their susceptibility. The reports indicate that mice and hamsters are the animals of choice, and mucin may enhance pathogenicity. In tissue sections the lesions appear primarily as abscesses with leucocytic infiltration containing sulfur granules with basically stained filaments and eosinophilic clubs.

*Actinomyces naeslundii.* Even though this species was established in 1950, pathogenicity studies have been minimal. Howell et al. (213) obtained minimal lesions in hamsters, but Buchanan and Pine (56) inoculated one strain into ten mice and obtained lesions of variable size in the peritoneal cavity of seven animals. Georg and Coleman (77, 141) inoculated 119 mice using twelve strains and obtained an 87% infectivity rate with more extensive lesions. It is now well established that *A. naeslundii* is consistantly found in human plaque and calculus, and recently Socransky et al. (438) isolated such a strain, inoculated gnotobiotic rats and demonstrated that the organism colonized the cervical and root surfaces of the teeth and formed destructive bacterial plaque leading to pocket formation, destruction of alveolar bone, root caries and exfoliation (Figure 11-18). These results have been confirmed by Frank et al. (127).

*Actinomyces eriksonii.* This strict anaerobe was described by Georg et al. (144) in 1965, and at that time four strains were injected intraperitoneally and subcutaneously or intravenously into mice which were examined at intervals up to 5 weeks. Of the thirty mice, none died, but a high percentage developed small intraperitoneal or subcutaneous abscesses but no progressive tissue invasion. The organisms were demonstrable on smears and could be cultured. The pathogenicity was considerably less than that demonstrated by *A. israelii* injected into other mice at the same time.

*Actinomyces odontolyticus.* The only animal experiments so far reported are those of Georg and Coleman (141) in which saline suspensions from BHIA slants were prepared from each of six strains and injected intraperitoneally into a total of sixty-eight mice. After 15 days, only six mice showed small abscesses on the peritoneal wall, which indicates that under these particular experimental conditions this species is the least pathogenic of the *Actinomyces* for mice.

*Actinomyces viscosus.* Four strains of hamster isolates were injected intraperitoneally into twenty-four mice (141), and twenty-three showed lesions; of these, eight animals had extensive lesions. Granules but not clubs were demonstrable. No such reports have been made on human isolates. Of greater interest and importance is the recent demonstration that *A. viscosus* can initiate periodontal disease. Jordan and Keyes (230) in 1964 isolated a filamentous organism (now identified as *A. viscosus*) from plaque of hamsters with naturally occurring periodontal disease. Cultures of the organism were inoculated orally and added to the drinking water of uninfected albino hamsters. The inoculated groups developed gingival accumulation of plaque with bone resorption. Other groups of hamsters inoculated with streptococci or other organisms were negative. These results have been confirmed using human isolates of *A. viscosus* in hamsters as well as in gnotobiotic rats (127, 232; see Figures 11-14 and 11-17).

*Arachnia propionica.* Buchanan and Pine (56) injected sixteen mice intraperitoneally with two strains, and abscesses were demonstrated in all animals. Later, Georg and Coleman (141) inoculated mice with two additional strains and reported that all mice developed significant lesions which resembled those produced by *A. israelii*.

# 12

# *Nocardia,* Nocardiosis and Nocardiomycosis

## HISTORY

In 1888 Nocard (342) isolated an aerobic filamentous organism from caseous lymph nodes of cattle with farcy on the island of Guadeloupe. He named the organism *Streptothrix farcinica.* Trevisan in 1889 (I. Genera e le Specie delle Batteriacee, Milano) applied the name *Nocardia farcinica.* There is controversy as to whether or not this should be the type species of the genus. Some authors (271) express the opinion that it should not because the original description is incomplete and verified cultures are not available.

Eppinger in 1891 (113) isolated an aerobic, gram-positive, acid-fast organism from a fatal case of meningitis with a brain abscess and named the organism *Cladothrix asteroides.* Finally, Blanchard (32) proposed the presently used name *Nocardia asteroides* (Eppinger) Blanchard. In 1909 Lindenberg (285) isolated another species from the leg of an Italian patient in Brazil and named the organism *Discomyces brasiliensis.* This was renamed *Nocardia brasiliensis* by Castellani and Chalmers in 1913 (68). In 1924 Snijders (434) isolated from the infected ear of a guinea pig an organism which he named *N. otitidis caviarum.* The organism was renamed *N. caviae* by Erikson (114). In this monograph we will consider only these three species which do cluster in numerical taxonomy studies (165, 258, 460).

## GENERAL CHARACTERISTICS

Nocardiae are aerobic, gram-positive, filamentous organisms (branching is demonstrable in slide cultures [138]) which are termed partially acid-fast because there is an irregular retention of carbol-fuchsin. Acid-fastness may be enhanced by growth on glycerol agar or in milk. The filaments fragment into bacillary and coccoid elements giving a very pleomorphic appearance. Colonies may be buff, orange, red, brown or white, soft or friable, smooth or granular and irregular, wrinkled or heaped. White aerial hyphae may be produced (Figures 12-1 and 12-2). The growth range is 10 C-50 C with 37 C optimal. Nocardiae produce acid from carbohydrates oxidatively and utilize such compounds as sugars, fatty acids, hydrocarbons and steroids as sole sources of carbon. Catabolism of these diverse compounds requires glycolysis, TCA cycle, hexose monophosphate pathway and $\beta$-oxidation. All of these have been reported in *Nocardia* (54, 67, 102, 318). Nocardiae are resistant to lysozyme. The organisms contain m-Dap, arabinose, galactose and nocardomycolic acid (carbon skeleton of about fifty carbons) in the cell wall (274, 308, 409). The G + C content of the DNA ranges from 67-69 moles %. Their normal habitat is the soil, and

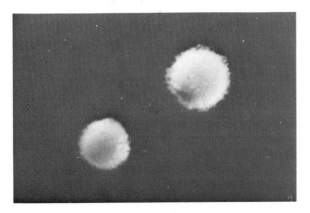

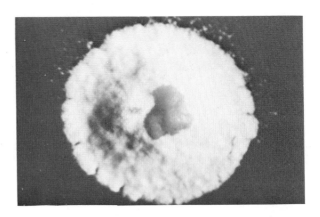

Figure 12-1.* *N. asteroides* colonies. BHI agar. Aerial filaments entirely covering the lower colony and very sparse on the upper colony. 37 C.

Figure 12-2. *N. asteroides* colony. Inorganic salts-yeast extract-glycerol agar in a humid atmosphere. 37 C.

all three species have been isolated from the soil using the paraffin-rod baiting technique (252, 255, 309). A nocardiophage has been described, but the susceptibility of these three species to phage has not been clearly demonstrated (55). See Table 12-1 for differential characteristics of the species.

Numerous authors have reported on the production of experimental infections in mice, guinea pigs, rabbits, hamsters, monkeys, dogs, calves and the chorioallantoic membrane of chick embryos (138, 295, 298, 306, 321, 361). In general, animal pathogenicity is: 1) strain variable, 2) enhanced by mucin, and 3) a localized response spontaneously healing or requiring large or repeated inocula to cause death. Clubbed granules have frequently been demonstrated with *N. brasiliensis* and *N. caviae* but not with *N. asteroides.* In recent reports on comparative pathogenicities, Gonzalez-Ochoa (164) produced a progressive infection in the foot pads of mice with *N. brasiliensis* but not with *N. asteroides* or *N. caviae.* Kurup et al. (256) used mice, guinea pigs, rabbits and the chorioallantoic membrane of chickens indicating that *N. caviae* was more virulent than the other two species. Uesaka et al. (461) reported that *N. asteroides* was more virulent than *N. brasiliensis* in mice, and then Smith and Hayward (430) indicated that *N. caviae* was more pathogenic than *N. asteroides* by the intravenous route in mice.

*Nocardia asteroides.* This species has the general characteristics given above but in comparison to the other two species it is less active biochemically. It can utilize a number of compounds as a carbon source but consistently produces acid only from glucose, does not hydrolyze casein or xanthine, is strain variable in the hydrolysis of hypoxanthine and tyrosine but does hydrolyze Tween 80. In relation to carbon sources it is interesting to note that paraffin-coated rods (Figure 12-4) have been used to isolate *N. asteroides* from soil and clinical specimens (320). Growth will usually occur at 45-50 C, and most strains will survive 60 C for 4 hrs. It should be stressed that the colonies of this organism vary considerably as to color (buff, orange, red), consistency (soft to friable) and surface texture (smooth to granular), and this is true even within a single strain (Figures 12-1, 12-2, 12-3). Some strains produce sufficient aerial hyphae to turn the colony white: others produce little or no aerial hyphae (Figures 12-1, 12-2). Cell-wall analysis was first done by Cummins et al. (91) and more recently by Lechevalier et al. (273). They (272) also have classified the aerobic actinomycetes into Types I-VIII on the basis of cell-wall composition with *Nocardia, Mycobacterium* and certain other genera in Type IV as they contain m-Dap, arabinose and galactose. These authors also use the presence of nocardomycolic acid (271, 274) as an identifying characteristic.

A number of authors have done immunological studies with whole cells, cell extracts, cell cytoplasm and culture filtrates as immunizing and/or test antigens using precipitin, gel diffusion, complement fixation, or hemagglutination tests (26, 121, 358, 402, 458, 486). Species specificity

*Figures 12-1, 12-2 and 12-3 show a single strain of *N. asteroides* grown with different carbon sources and with or without increased humidity, demonstrating the colony variation occurring under different environmental conditions.

 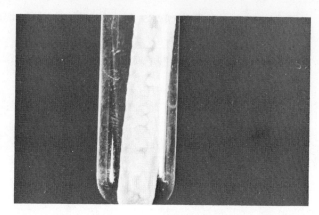

**Figure 12-3.** *N. asteroides* colony. Inorganic salts-yeast extract-glucose agar in a humid atmosphere. 37 C.

**Figure 12-4.** *N. asteroides* growing along a paraffin rod in Czapek's broth.

has not been clearly established, and Kwapinski et al. (260) indicate cross-reactivity with mycobacteria. Pier and Fichtner (359), using soluble extracellular antigens and the gel diffusion test, designate four distinct serological types of *N. asteroides* with only Type IV showing cross-reactivity with other species. Al-Doory (6) successfully applied the fluorescent antibody technique to identify *Nocardia* but was unable to entirely eliminate cross-reactivity among the three species.

Drake and Henrici (100) demonstrated that guinea pigs could be sensitized with *N. asteroides,* and since that time there have been continuing attempts to develop a skin test antigen which would aid in the diagnosis of nocardiosis and distinguish particularly between nocardiosis and tuberculosis. Such preparations have included: 1) sensitins prepared in the general manner of tuberculins (4, 35, 300), 2) ethanol precipitates of culture filtrates (106), 3) trichloracetic acid extracts of filtrates (360), 4) a phenol-extracted polypeptide (245, 246) and 5) cytoplasmic extracts (347). As yet, however, a completely satisfactory antigen which will not cross-react with other species of *Nocardia, Mycobacterium* or *Streptomyces* and which has been adequately tested in man has not been developed.

It has been well established that *N. asteroides* is usually sensitive to the sulfonamides but not responsive to most antibiotics, including rifampin used for treating other bacterial, mycobacterial or mycotic infections. There have been a number of reports on *in vitro* testing of the susceptibility of *N. asteroides* to antibiotics or other chemotherapeutic agents (30, 31, 346, 450). Bach et al. (12) using some forty-five antimicrobial agents have shown that a number of strains were resistant to sulfonamides and that the most active antibiotic was minocycline, a tetracycline derivative. As a guide to therapy, they stress the importance of determining the resistance or susceptibility of the isolate, and this suggestion is supported by the work of Lerner and Baum (281).

*Nocardia brasiliensis.* This species is aerobic, filamentous, partially acid-fast and in culture often fragments into coccobacilli. It clusters as a species in numerical taxonomy studies (165, 258, 460). As with *N. asteroides,* there is variation in colony appearance in that the color may be yellow to yellowish orange or buff or even white if aerial hyphae are formed. The colonial consistency varies from dry and friable to moist and soft with a rough to slightly irregular surface. The general comparative characteristics of this species are listed in Table 12-1. *N. brasiliensis* is differentiated from the other species by hydrolyzing casein, tyrosine and urea, utilizing citrate as a carbon source, producing acid from galactose and mannose and not growing above 45 C. Bojalil and Cerbon (35) describe a procedure using 0.4% gelatin in which *N. brasiliensis* produces characteristic colonies with a positive ninhydrin test. The G + C content of the DNA is 67-68 moles %. The cell wall contains m-Dap, arabinose, galactose and nocardomycolic acid (271). Zamora et al. (486) isolated a species-specific polysaccharide, and Bojalil and Magnusson (36) prepared a sensitin which gave positive skin tests in *N. brasiliensis* mycetoma patients and negative reactions in patients with tuberculosis or mycetomas due to other organisms. The normal habitat of this bacterium is the soil, and this is the source of infection in man (343).

**Table 12-1. Comparative characteristics of *Nocardia* and *Streptomyces*[a,b].**

| | *Nocardia asteroides*[c] | *Nocardia brasiliensis* | *Nocardia caviae* | *Streptomyces madurae*[a] | *Streptomyces pelletieri*[a] |
|---|---|---|---|---|---|
| Gram | + | + | + | + | + |
| Acid-fast[d] | + | + | + | − | − |
| Catalase | + | + | + | + | + |
| **Growth at:** | | | | | |
| 10 C [e] | −(+) | +(−) | −(+) | +(−) | +(−) |
| 50 C | +(−) | − | −(+) | − | − |
| **Carbon sources:** | | | | | |
| Citrate | −(+) | + | −(+) | −(+) | − |
| Malate | +(−) | + | + | +(−) | − |
| Paraffin | + | + | + | −(+) | − |
| Succinate | + | + | + | + | − |
| **Acid from:** | | | | | |
| Arabinose | − | − | − | + | − |
| Galactose | −(+) | + | − | +(−) | − |
| Glucose | +(−) | + | + | + | + |
| Inositol | −(+) | + | + | +(−) | − |
| Mannitol | − | + | +(−) | + | − |
| Mannose | −(+) | +(−) | −(+) | + | − |
| Trehalose | +(−) | + | +(−) | +(−) | + |
| Xylose | −(+) | − | − | + | − |
| **Hydrolysis of:** | | | | | |
| Casein | − | + | − | + | + |
| Hypoxanthine | −(+) | +(−) | + | +(−) | +(−) |
| Tyrosine | −(+) | + | − | +(−) | + |
| Xanthine | − | − | + | − | −(+) |
| Urea | +(−) | + | + | − | − |
| **Cell Wall[f]** | | | | | |
| Meso-DAP | + | + | + | + | + |
| LL-DAP | − | − | − | − | − |
| **Sugars:** | | | | | |
| Arabinose | + | + | + | − | − |
| Galactose | + | + | + | + | + |
| Madurose[g] | − | − | − | + | + |

a The generic names *Nocardia* and *Actinomadura* are also used for *S. madurae* and *S. pelletieri*.

b For differentiation of other *Streptomyces*, *Micromonospora* and Rhodochrous Group see Berd (23).

c Berd (21) divides this species into two groups designated as acetamidase + and acetamidase − with other associated characteristics.

d Kinyoun's carbol-fuchsin for 3 min. at room temperature, decolorize with 1% sulfuric acid; staining varies from good to partial to negative.

e Results are frequently variable and difficult to duplicate.

f Using the methods of Becker et al. (20), Lechevalier (270) and Berd (23).

g Madurose is 3-0-methyl-D-galactose (274).

*Nocardia caviae (N. otitidis caviarum).* This species was not well defined until the work of Gordon and Mihm (170) but is now an established species which clusters in numerical taxonomy studies (165, 258, 460). This bacterium morphologically resembles the other nocardiae in being partially acid-fast, pleomorphic and forming yellowish or buff or white colonies. The biochemical reactions are listed in Table 12-1. Those tests of differential value are: hydrolysis of xanthine, nonutilization of citrate as a sole carbon source and acid production from mannitol but not from galactose. Erythrocyte hemolysis has been reported (430) but is not common for the nocardiae. It has m-Dap, arabinose, galactose and nocardomycolic acid in the cell wall. The normal habitat is the soil (255), but it has been isolated from a dog (242). *N. caviae* is sensitive to the sulfonamides (31), but strain variation is to be expected. The G + C content of the DNA is 65 moles %. *N. Caviae* can be the etiological agent of mycetoma in man (452), but its role in disseminated or pulmonary infections has not been established. See the section below for a discussion of mycetomas.

*Nocardiosis.* The nocardiae cause three clinical manifestations: systemic nocardiosis, mycetoma and a lymphocutaneous syndrome. Systemic involvement is due almost entirely to *N. asteroides;* either *N. brasiliensis* or *N. caviae* cause mycetomas; and *N. brasiliensis* causes the more recently described lymphocutaneous syndrome which clinically may resemble sporotrichosis.

Nocardiosis usually begins as a pulmonary infection which may be transitory, acute or chronic, localized or disseminated and clinically may simulate bronchopneumonia, tuberculosis, lung abscess or neoplasm (2, 166, 177). Pathologically, the infection is characterized by acute necrosis with abscess formation. About 30% (275) remain pulmonary, but metastasis can occur to any tissue or organ of the body with some predilection for the central nervous system, particularly the brain (Figures 12-5 and 12-6). Here, single or multiple abscesses may develop, but extension to the spinal cord is not common, and the spinal fluid is usually negative for smears and cultures (414, 464, 474, 475). Hoeprich et al. (199) reviewed 148 cases of nocardiosis in which forty-five (30%) had cerebral involvement with an 85% mortality rate. Next most frequent sites of metastasis were the spleen, kidney, heart, liver and bones (329). Primary extrapulmonary infections of the eye (16), arm (305), intestine (64) and even dissemination (377) have been reported.

Three surveys indicate that approximately 390 cases of nocardiosis have been reported through 1967. An analysis by Murray et al. (329) of 179 cases, mostly from Europe and the U.S., indi-

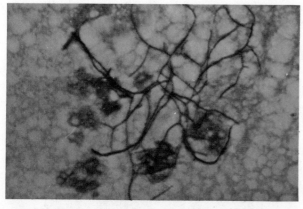

Figure 12-5. *N. asteroides* in direct smear from cerebral abscess of fatal case of nocardiosis. The filaments were acid-fast.

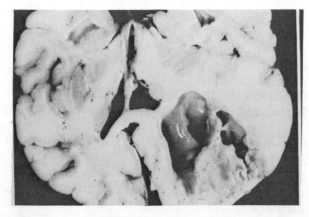

Figure 12-6. Cerebral abscess from a fatal case of nocardiosis.

cated that 70% were males who ranged in age from 4 mos. to 87 yrs., with the majority between 20 and 60 yrs. old. The mortality rate prior to the use of sulfonamides was 75% and since then 46%. Susens et al. (446) listed 174 cases occurring between 1960-1967, with an organ involvement of ninety-two lung, thirty-seven brain, eight pericardium, five kidney, five bone and three eye, with a mortality rate of 56%. This approximate doubling of reported cases in 7 years supports the thesis that the incidence of nocardiosis is increasing. Kurup et al. (256) gives references for 292 cases from 1944-1968 with the majority from the U.S., but other references from a variety of countries demonstrate the worldwide occurrence of nocardiosis. A number of authors report that *N. asteroides* is an opportunist and that the incidence of infections is increasing in patients with leukemia, Hodgkin's disease, lymphosarcoma, renal transplants or who are receiving corticosteroids or immunosuppressive therapy (177, 352, 484).

The habitat of *N. asteroides* is the soil (168, 252, 309), and it is assumed that this is the source of infection in man. It is not a normal inhabitant of the oral cavity of man. Raich et al. (381) in over 50,000 sputum samples isolated nocardiae only seven times, and Kurup et al. (254) cultured sputa from 235 patients with bronchopulmonary disease using the paraffin-bait technique and isolated *N. asteroides* from only one patient which was then diagnosed as nocardiosis. Therefore, if *N. asteroides* is isolated from sputum, nocardiosis should be considered even without clinical evidence, and some authors advocate the initiation of therapy.

*N. asteroides* causes infections in cats, goats, pigs, horses, dogs, monkeys, rodents, chickens and rainbow trout and severe mastitis in cattle (121, 224, 242, 249, 361, 433), but there is no indication of direct transmission from animals to man nor for that matter from man to man.

In 1943 Lyons et al. (293) reported the first successful treatment of a case of nocardiosis with a sulfonamide. The sulfonamides have remained the drug of choice, particularly sulfadiazine with dosages of 4-6 g/day for a prolonged period. Other antimicrobials which have been used with or without sulfonamides include cycloserine, trimethoprin, ampicillin, streptomycin and minocycline (12, 74, 199, 330, 346, 464, 475). Recently, Lerner and Baum (281) reported on *in vitro* susceptibility testing which emphasized strain variability and supported the concept that each isolate should be tested against a variety of antimicrobials with attention to inoculum size and pH as well as to time and temperature of incubation.

*Mycetoma or Nocardiomycosis.* *N. brasiliensis* is one of the etiological agents of mycetoma with the organisms most likely being introduced into the skin by thorns, splinters or an abrasion. This usually involves the lower extremities as a chronic localized infection of the subcutaneous tissue forming granulomatous masses with the development of abscesses and draining sinuses. The pus usually contains demonstrable granules which may be white, yellow, red, brown or black in color. Gonzalez-Ochoa (163) analyzed 103 cases indicating that 62% involved the lower extremities, 25% the trunk, 10% the upper extremities and 2% the head and neck. The frequency of feet and back infections is said to relate to the soil habitat of the organism. Symptoms may occur within days or months of contact with the organism. The infection may invade contiguous tissue, and, although hematogenous spread is not common, it can involve any tissue including bone, joints, lung, thorax or abdomen. Recently, a *lymphocutaneous syndrome* resembling sporotrichosis has been described (350), and in addition five cases of systemic nocardiosis due to *N. brasiliensis* have been reported in the U.S. (69). Men are infected three times as frequently as women, and most cases occur between the ages of 20 and 40, although no age is exempt. Sulfonamides are the drug of choice, although dapsone and trimethoprin (301) have been used. Lerner and Baum (281) report *in vitro* sensitivity of some strains to erythromycin and gentamicin.

The majority of cases have been reported from Africa, India, Mexico and South America with infrequent reports from Australia, the Caribbean, Germany, Italy, Japan, Portugal and the United States. Boyd and Crutchfield in 1921 (44) reported the first mycetoma case in the United States, and from then through 1972 a total of sixty-nine cases have been listed (22, 200). An additional twenty-six cases of *N. brasiliensis* infections have been reviewed by Berd (22). This organism is the most common cause of mycetoma in Mexico and Central and South America, but in countries with a higher incidence of mycetoma, such as Africa and India, other organisms are more frequently involved, including: *Streptomyces madurae, S. pelletieri, S. somaliensis, Madurella mycetomi, M. grisea, Monosporium apiopermum, Cephalosporum* sp. and *Phialaphora jeanselmi.*

*N. asteroides, N. caviae* and *A. israelii* have been infrequently reported as well as the rare instance of botryomycosis caused by *Staphylococcus aureus.*  D. J. Winslow has a good discussion of this entire subject in the book edited by R. D. Baker (14).

In addition to the clinical picture, the specific diagnosis depends upon:  1) demonstration of granules in the pus or in tissue sections, 2) demonstration of the organism in smears from the exudate, and 3) isolation and identification of the organism.  Grossly, the granules vary in color from white, yellow, red, brown to black, and there is a degree (probably not high) of relationship between the color and probable etiological agent; e.g., red granules suggest *S. pelletieri* and brown suggest *Madurella.*  An initial wet-mount examination should indicate whether or not fungal hyphae or the fine filaments of *Nocardia* or *Streptomyces* are present.  For a more definitive differentiation of these organisms, smears or sections can be stained by gram's, acid-fast, periodic acid-Schiff or Gomori methenamine-silver methods. For specific identification, the organism must be isolated and characterized.  The granules of *N. brasiliensis* are small, usually less than 1 mm, irregular, spherical and yellowish in color.  (For ultrastructure morphology see reference 297.)
In smears, the filaments are gram-positive and acid-fast to the degree that some filaments are red, others partially stained and some not stained at all.  Theoretically, this should separate *Nocardia* and *Streptomyces*, but the distinction is not always clear-cut.  The procedures for culturing are given in the next section.

## LABORATORY PROCEDURES FOR *NOCARDIA*

**Introduction.**  In addition to the clinical picture, a definitive diagnosis depends upon the demonstration of the organisms in smears (Figure 12-5) or sections along with the isolation and identification of the nocardiae.

**Materials.**  Sputum, bronchial washings, pus, exudate or spinal fluid are the materials usually received in the laboratory.  The sputum should not be digested even though *Nocardia* may survive the procedure.  If possible, spread the material out in a petri dish and observe for clumps of the organisms which may resemble granules.  *N. asteroides* may grow in masses, but club formation is indeed rare.  *N. brasiliensis* and *N. caviae* will form white or yellow granules with clubs.

**Microscopic.**  If clumps or granules are present, selectively remove and crush between slides.  Examine some of the material as a wet mount to be certain that fungal hyphae are not present.  Prepare duplicate smears for staining.  Stain one smear by gram's and the other with Kinyoun's carbol-fuchsin for 3 min. at room temperature and decolorize with 1% $H_2SO_4$ (138). The presence of gram-positive filaments with some degree of acid-fast staining is an indication of *Nocardia.*  The acid-fast stain may vary from good to variable to negative and is often difficult to interpret.  Positive and negative controls should be included.

**Culture.**  Clumps or granules, if present, should be washed at least twice in sterile saline and then crushed.  Streak these or any of the other material onto duplicate plates of Sabouraud dextrose agar plus duplicate plates of either BHIA or blood agar (nocardiae will grow on Lowenstein's).  Incubate one plate of each pair at 37 C and the other at 25 C.  Observe at frequent intervals for 3 weeks, but detectable growth usually occurs within 1 week.  Colony descriptions are given above in the discussion of each species.

The final identification depends upon the results of a battery of tests including utilization of various carbon sources, hydrolysis of different substrates, acid formation from carbohydrates and cell-wall analysis from whole-cell hydrolysates.  The results from such tests are recorded in Table 12-1.  Procedural details of each test are beyond the scope of this monograph but are adequately described by Berd (21, 23), Georg et al. (138) and Gordon and Mihm (170).  Important but less frequently done procedures include the determination or demonstration of:
1. *Oxidative reactions on sugar* by following the general techniques of Hugh and Liefson (218) in which tubes of glucose broth are inoculated and incubated aerobically and anaerobically.
2. *Branching filaments* of the microcolonies in slide cultures (138).
3. *Meso-Dap, arabinose* and *galactose* in the cell wall using whole-cell hydrolysates (20, 23, 270, 274).

4. *Nocardomycolic acid* by paper chromatography as detailed by Lechevalier et al. (271). An extraction method giving quantitative differences between *Mycobacterium* and *Nocardia* is described by Kanetsuna and Bantoli (235).

# 13

# Practical Procedures for Isolation and Identification

## INTRODUCTION

The presence of *Actinomyces* in clinical specimens is usually suggested by a tentative diagnosis of actinomycosis or by the source of the specimen. The specimens usually consists of pus, exudates from draining sinuses or sputum. Pus and sputum should be examined for the presence of yellow or whitish granules, but granules are not always present. In addition, in any clinical specimen, *Actinomyces* and *Arachnia* should always be considered whenever direct smears from this material show gram-positive, non-acid-fast rods in diphtheroidal arrangements with or without the presence of branching filaments.

Of the organisms discussed in this monograph, *Actinomyces*, *Arachnia*, *Rothia* and *Nocardia* can be isolated from clinical materials. These organisms must be differentiated from each other and from *Propionibacterium*, *Eubacterium*, *Bifidobacterium* and *Lactobacillus*, which occur in the same materials. Laboratory procedures for *Nocardia* were included in Chapter 12 and will not be discussed here except to indicate *Nocardia*'s differentiation from the other genera. So far as is known, *Bacterionema* occurs only as a part of the normal flora where it must be considered in studies involving dental plaque and calculus. Since this discussion is concerned with isolation from clinical materials rather than with normal flora studies, *Bacterionema* has not been included.

The following protocol is designed to aid in isolation and identification of *Actinomyces*, *Arachnia* and *Rothia* using a minimum of tests. In all cases, the appropriate chapter in the monograph will give detailed information and additional tests if these are needed in specific cases. A flow chart outlining isolation and identification is shown in Figure 13-1.

## ISOLATION

In examinations for granules, pus and sputum should be poured into a sterile petri dish and examined for yellowish to white granules which are hard in consistency. Select a granule, place it on a glass slide and crush under a coverslip for microscopic examination.

## INOCULATION OF PRIMARY CULTURES

If granules are present, wash a granule in two or three changes of sterile saline in a petri dish. Crush the granule in a small amount of saline using a sterile glass rod, and use the resulting

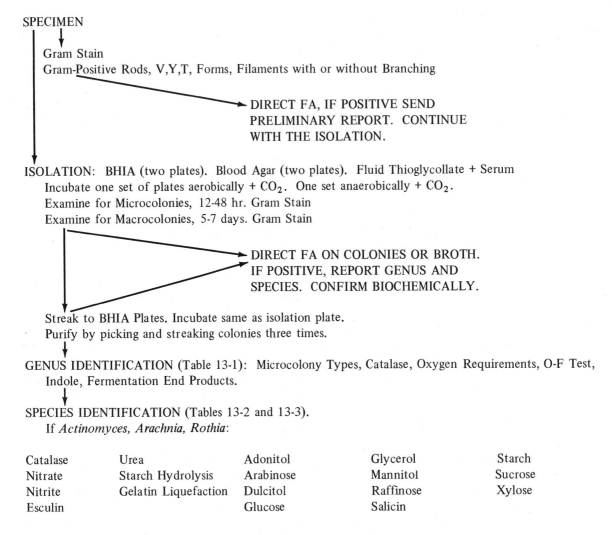

SPECIMEN

Gram Stain
Gram-Positive Rods, V,Y,T, Forms, Filaments with or without Branching

DIRECT FA, IF POSITIVE SEND
PRELIMINARY REPORT. CONTINUE
WITH THE ISOLATION.

ISOLATION: BHIA (two plates). Blood Agar (two plates). Fluid Thioglycollate + Serum
Incubate one set of plates aerobically + $CO_2$. One set anaerobically + $CO_2$.
Examine for Microcolonies, 12-48 hr. Gram Stain
Examine for Macrocolonies, 5-7 days. Gram Stain

DIRECT FA ON COLONIES OR BROTH.
IF POSITIVE, REPORT GENUS AND
SPECIES. CONFIRM BIOCHEMICALLY.

Streak to BHIA Plates. Incubate same as isolation plate.
Purify by picking and streaking colonies three times.

GENUS IDENTIFICATION (Table 13-1): Microcolony Types, Catalase, Oxygen Requirements, O-F Test,
Indole, Fermentation End Products.

SPECIES IDENTIFICATION (Tables 13-2 and 13-3).
If *Actinomyces, Arachnia, Rothia*:

| Catalase | Urea | Adonitol | Glycerol | Starch |
| Nitrate | Starch Hydrolysis | Arabinose | Mannitol | Sucrose |
| Nitrite | Gelatin Liquefaction | Dulcitol | Raffinose | Xylose |
| Esculin | | Glucose | Salicin | |

suspension to inoculate media. If granules are not present, inoculate cultures with well-mixed pus or exudate. If the specimen is sputum, any small concretions or obviously purulent flecks should be selected for culture.

Inoculate the following media using about 0.1 ml inoculum for each:
1. Fluid Thioglycollate Broth (BBL or Difco) containing 0.1 to 0.2 % sterile rabbit serum.
2. Two Brain Heart Infusion Agar plates.
3. Two blood agar plates (use only animal blood).

Incubate one BHIA and one blood plate aerobically with $CO_2$ and the other pair anaerobically with $CO_2$. A Gas-Pak or Torbal anaerobic jar with a freshly reactivated catalyst is satisfactory.

## EXAMINATION OF PRIMARY CULTURES

Examine BHIA plates at 12-24 hrs., using a microscope at 100X, for the presence of tiny, filamentous microcolonies. If filamentous microcolonies are not seen, reincubate plates and re-examine at 48 hrs. If filamentous microcolonies are seen, cut out the section of agar containing the colony (this may be done with an inoculating needle, flattened and sharpened on one edge) and place the agar section in a tube of fluid thioglycollate broth with serum. Reincubate plate for 5-7 days.

Using a stereoscopic microscope, examine BHIA and blood agar plates after 5-7 days for macrocolonies. *A. israelii* and *A. propionica* usually produce nodular, molar-tooth or raspberrylike colonies. However, other species and some strains of these species produce smooth colonies which are not particularly distinctive. A finely granular or ground-glass surface on a smooth, convex or umbonate colony is suggestive of *Actinomyces*. Make a gram stain of at least one colony of each colony type on the plate. Colonies which contain gram-positive diphtheroidal rods should be picked to thioglycollate broth with serum and to another plate.

After 48 hrs. and again at 5 and 7 days, remove some of the growth (be certain to include growth and not clear broth) from the primary fluid thioglycollate broth culture using a Pasteur pipette, smear and gram stain. If gram-positive diphtheroids or filaments are seen, subculture 0.1 ml onto each of two BHIA and two blood agar plates. Incubate and examine plates as above.

## PURIFICATION OF ISOLATES

Pick colonies or growth in thioglycollate and streak onto BHIA plates. Incubate under the same conditions as the isolation plates. To *purify*, pick isolated colonies to BHIA plates—one plate can be sectioned to accommodate four to six colonies—and incubate. Repeat the picking and streaking of isolated colonies at least three times. This procedure cannot be overemphasized as it is extremely easy to transfer cocci or diphtheroids along with the actinomycete, and this is not detected until a later date. It is sometimes difficult to subculture microcolonies so it may be necessary to use larger well-isolated colonies.

## IDENTIFICATION OF ISOLATES

The identification of both the genus and species of these organisms is based on morphology, biochemical tests and serology (particularly FA).

Identification at the genus level is based on gram-stain morphology, colony morphology, oxygen requirements and a few selected biochemical tests including the analysis of end products of glucose fermentation. Gram-stain morphology is the initial criterion for considering that an organism belongs to one of the genera under consideration. These bacteria are gram-positive rods with Y, V, T forms and Chinese letter arrangements. Long filaments with or without branching may be present. Slender, irregularly staining organisms of this type usually belong to the genera *Actinomyces, Arachnia, Rothia, Corynebacterium* or *Propionibacterium*. Evenly stained, thick rods with bifurcate or clubbed ends are usually *Eubacterium, Bifidobacterium* or *Lactobacillus*. Identification to the genus level may be sufficient for most purposes. Laboratories which are not equipped for detailed study should submit cultures to a reference laboratory for speciation.

The tests shown in Table 13-1 can be used to indicate the genus to which an organism with these morphological characteristics belongs. If information concerning end products of glucose fermentation is available, most organisms can be placed in the proper genus on the basis of this information and the catalase test. If this information is not available, the other tests listed will usually suffice to eliminate organisms of the genera *Bifidobacterium, Propionibacterium, Corynebacterium* and *Lactobacillus* from further consideration. *Nocardia* is aerobic, oxidative, grows with an inorganic nitrogen source and is usually partially acid-fast.

Organisms which seem to belong to the genera *Actinomyces, Arachnia* or *Rothia* are then subjected to the tests shown in Tables 13-2 and 13-3 for speciation. Inoculate a tube of BHI broth and incubate 2-3 days as an inoculum for the biochemical tests. Homogenize the growth on a Vortex mixer or some similar device and, if necessary, dilute the culture with fresh sterile broth until it is barely turbid. Cultures which grow in hard granules emulsify more satisfactorily if a few glass beads are added to the tube before vortexing. With such cultures, care must be taken that the suspension does not settle out in the pipette so that all biochemical tests get an even inoculum. Inoculate each biochemical test with 0.1 ml of the suspension and incubate under optimal conditions as determined by an oxygen-requirements test. If this information is not available, catalase-negative strains will almost always grow well anaerobically with $CO_2$. Catalase-

Table 13-1. Identification of genera of gram-positive diphtheroids[a].

| End Products of Glucose Fermentation | Succinic Lactic Acetic Formic (tr) | Succinic Lactic Acetic Formic (tr) | Succinic Lactic Acetic Formic (tr) | Propionic Acetic Formic | Propionic Acetic Formic | Propionic Lactic Acetic 1.5:1 | Lactic | Butyric and Others | Lactic Acetic Succinic (tr) |
|---|---|---|---|---|---|---|---|---|---|
| Microcolony | Fil. | Fil. | S | Fil. | S | $S^{fil}$ | S | S | $S^{fil}$ |
| Catalase | − | + | − | − | + | − | − | − | + |
| Oxygen Requirement | Fac, An | Fac. | Fac., An | Fac. | $An^{fac}$ | An | An | An | Aer. |
| O-F Test | F | F | F | F | F | F | F | F | F |
| Indole | − | − | − | − | + | − | − | −+ | − |
| ACTINOMYCES | + | + | | | | | | | |
| ARACHNIA | | | + | | | | | | |
| ROTHIA | | | | | | | | | + |
| PROPIONIBACTERIUM | | | | | + | | | | |
| EUBACTERIUM | | | | | | | | + | |
| BIFIDOBACTERIUM | | | | | | + | | | |
| LACTOBACILLUS | | | | | | | + | | |

a *Nocardia* is oxidative (no end products), acid-fast, grows on inorganic nitrogen, catalase + and aerobic.

**Table 13-2.  Identification of Arachnia and Actinomyces species.**

| | A. bovis | A. odontolyticus | A. viscosus | A. naeslundii | A. israelii | A. propionica | P. acnes | C. pyogenes |
|---|---|---|---|---|---|---|---|---|
| Catalase | –ᵃ | – | + | – | – | – | +⁻ | – |
| Nitrate | – | + | + | + | V | + | + | – |
| Nitrite | – –ᵇ | – | V | V⁻ | – | – | – | – |
| Esculin | + | V⁺ | + | + | + | – | – | – |
| Urea Agar | – | – | + | + | – | – | – | – |
| Urea Broth | – | – | – | – | – | – | – | – |
| Starch Hydrol. (wide) | + | – | – | – | – | – | – | – |
| Gelatin Liq. | – | – | – | – | – | – | + | + |
| Adonitolᶜ | – | – | – | – | Type 1  V⁺ | + | V | V⁻ |
| Arabinose | – | V⁻ | – | – | Type 2  – | – | – | V⁻ |
| | | | | | + | | | |
| Dulcitol | + | – | – | – | + | – | – | – |
| Glucose | + | + | + | + | + | + | + | + |
| Glycerol | V | V | + | + | – | – | +⁻ | V |
| Mannitol | – | – | – | – | V⁺ | + | V | + |
| Raffinose | – | V⁻ | + | + | + | + | – | – |
| Salicin | V | V | V⁺ | V⁺ | + | – | – | – |
| Starch | + | + | V | V | V | V⁺ | – | + |
| Sucrose | V⁺ | + | + | + | + | + | – | + |
| Xylose | V | V⁺ | – | – | + | – | – | + |

ᵃ  – = 90% or more negative
    + = 90% or more positive
    V⁺ = variable usually + (65–90% +)
    V = variable (30–65% +)
    V⁻ = variable usually – (10–30% +)
    Results are based on tests read at 7 days except sugars.

ᵇ Information not available.

ᶜ Based on tests done in thioglycollate fermentation base with final reading at 14 days.

**Table 13-3. Identification of *Rothia*.**

|  | *Rothia dentocariosa* | *Rothia* sp. |
|---|---|---|
| Catalase | +[a] | + |
| Nitrate | + | + |
| Nitrite | V+ | V |
| Urease—Thioglycollate | —[b] | V— |
| | | |
| Arabinose | —[c] | — |
| Glucose | + | + |
| Glycerol | V+ | + |
| Lactose | V— | V |
| Mannose | V+ | + |
| Mannitol | — | V+ |
| Raffinose | — | V+ |
| Rhamnose | V— | + |
| Salicin | V+ | + |
| Sucrose | + | + |
| Trehalose | V+ | + |
| Xylose | — | — |

[a]   — = 90% or more negative
  + = 90% or more positive
  V+ = variable, usually + (70-90% +)
  V  = variable
  V— = variable, usually — (10-30% +)

Results are based on tests read at 7 days except sugars.

[b] Biotype 3 strains are urease +.

[c] Based on tests done in meat-extract peptone medium with final reading at 14 days. Biotypes vary in certain sugars especially lactose, mannose, salicin and trehalose. See Tables 8-1 and 8-2, pp. 75-76.

positive strains (*Rothia* and *A. viscosus*) should be incubated aerobically with $CO_2$. For poorly growing isolates, add 0.1 to 0.2 ml sterile animal serum to all biochemical test media. Thioglycollate Fermentation Base is recommended for *Actinomyces* and *Arachnia,* and Meat Extract Peptone Fermentation Base is recommended for *Rothia.*

Most strains of *Actinomyces* and *Arachnia* can be speciated using the tests shown in Table 13-2. The reactions of *P. acnes* have been included because this organism not only occurs in clinical material but often occurs as a contaminant in previously purified cultures. *Corynebacterium pyogenes* is included because it must be differentiated from *Actinomyces* especially when dealing with material from animals since it resembles *A. bovis* in many ways.

## SEROLOGY

The Fluorescent Antibody (FA) procedure is the most rapid and reliable means of identifying *Actinomyces, Arachnia* and *Rothia* when FITC-conjugated antisera are available.

This technique can be used with any type of material in which the organism is present as listed below. This should always be followed by the isolation and identification of the organism by cultural means, although in some instances where a culture will not grow or the tissue has been fixed, FA offers the only means of specifically identifying the organism. Materials in which we have identified these organisms using direct FA include:

1.  Clinical material in which gram-positive diphtheroids or filaments were demonstrable.  Such material may contain a variety of different bacteria as well as two or three species of *Actinomyces*.
2.  Formalin-fixed tissue or histological sections.
3.  Gram-positive diphtheroids or filaments present on primary isolation plates or in thioglycollate broth.  This included cultures mixed with other bacteria.
4.  Pure culture isolates from plates or growing in broth.

# 14

# Techniques, Media
# and Reagents

Most of the media used for studying *Actinomyces* are available from commercial sources and are prepared and used according to the manufacturers' instructions. Details of modifications of commercial media and formulas for media not available commercially will be given in the following section.

## GENERAL-PURPOSE CULTURE MEDIA

1. Fluid Thioglycollate Broth (THIO) supplement with 0.1-0.2% sterile rabbit serum as needed
2. Brain Heart Infusion Broth (BHI)
3. Trypticase Soy Broth (TSB)
4. Brain Heart Infusion Agar (BHIA)
5. Trypticase Soy Agar (TSA)
6. Blood Agar-Heart Infusion Agar or TSA with 5.0% defibrinated rabbit blood. Sheep or horse blood generally gives comparable results.

## EXAMINATION OF COLONY CHARACTERISTICS

**Microcolonies.** *Medium:* BHIA plates with no more than 10 ml medium per plate.
*Procedure:* Streak BHIA plate for isolation; incubate and examine at 18-24 hr. The colonies may be observed directly on the medium within the plate or a small section of the agar may be removed and placed on a glass slide. Observe at 100X to 400X magnification. The lower magnification is useful for routine examination, but higher magnification may be necessary for detailed study of a colony. For observation under oil immersion, place a sterile coverslip over the colony. A 54X oil immersion objective is very useful for this purpose.
**Mature colonies.** *Medium:* BHIA and blood agar plates.
*Procedure:* Streak BHIA and blood agar plates for isolation. Incubate and observe after 7 and 14 days using a dissecting microscope.

## TESTS FOR OXYGEN REQUIREMENTS

1. **Agar Slant Method.** *Medium:* BHIA slants with cotton plugs.

*Inoculum:* Inoculate BHI with each test culture and incubate 2-3 days. Centrifuge and re-suspend growth in 0.85% saline to an O.D. reading of 0.5 on a BBL Spectronic 20 or to match a MacFarland #3 standard. Alternatively, the culture may be mixed thoroughly and diluted to match this turbidity standard with sterile broth.

*Procedure:* Inoculate eight BHIA slants per culture using a capillary pipette. Place the tip of the pipette containing the well-mixed suspension at the bottom of the slant, force out one small drop, and draw the pipette in a single line from the base to the top of the slant. Incubate slants, in duplicate, under the different conditions as described below.

    a. For aerobic incubation, place two slants with the original cotton plugs in the incubator.

    b. For aerobic + $CO_2$, anaerobic + $CO_2$ and anaerobic incubation, handle the tubes as follows: Clip off the cotton plug, push this short plug into the tube to just above the slant and place a small pledget of absorbent cotton on top of the plug. Add reagents to the absorbent cotton as listed below and immediately close the tube with a rubber stopper.

Aerobic + $CO_2$    – 5 drops 10% $Na_2CO_3$ + 5 drops lM $KH_2PO_4$

Anaerobic         – 5 drops 10% KOH + 5 drops pyrogallol solution (100 g pyrogallic acid in 150 ml distilled water)

Anaerobic + $CO_2$ – 5 drops 10% $Na_2CO_3$ + 5 drops pyrogallol solution

Incubate and record results after 3 and 7 days.
      3+ — best growth under any of the above conditions
      2+ — medium growth in comparison to the best
      1+ — light growth in comparison to the best
      0   — no growth

If the results on the duplicate tubes do not agree, the test should be repeated.

    **2. Agar Deep Method**. *Medium:* Trypticase Soy Agar + 0.2% glucose, distributed in deep tubes to a depth of at least 65 mm.

*Procedure:* Inoculate melted and cooled medium while still liquid. Using a capillary pipette, push the tip of the pipette to the bottom of the agar and withdraw it slowly while expelling a drop of inoculum. Mix the agar by rotating tube gently and let stand upright until hard. Incubate and read at 3 and 7 days. Measure the zone of maximum growth in mm down from the surface. Total growth zones may be recorded, but maximum growth zones give better differentiation. If results in the two tubes are not similar, repeat the test. Gas is not produced by the actinomycetes which differentiate these organisms from some *Bifidobacterium* and *Eubacterium*. If 1.0% glucose is added to the above medium, it can be used both for the determination of oxygen requirements and the production of gas.

## INOCULATION AND INCUBATION OF BIOCHEMICAL TESTS

**Inoculum.** Inoculate a tube of BHI and incubate 2-3 days. Mix thoroughly on a Vortex mixer or similar mixing device. A few glass beads in the tube will help to disperse very granular cultures. If necessary, adjust the suspension so that it is barely turbid. Use approximately 0.1 ml per tube of the resulting suspension to inoculate all biochemical tests except where special instructions are given.

**Incubation.** All tests are incubated at 35-37 C and in general are observed at 3, 7 and 14 days. When possible, tests should be incubated under optimal conditions as determined by oxygen requirement tests. If this information is not available, most catalase-negative cultures do best when incubated anaerobically with added carbon dioxide. Use a Torbal anaerobic jar with a gas mixture of $N_2$ – $H_2$ – $CO_2$ (80:10:10%) or a Gas-Pak jar with a $CO_2$ type generating packet. Simple gas exchange jars flushed with a mixture of $N_2$ – $CO_2$ (95:5%) may be used but are generally less satisfactory than catalyst-type jars. Catalase-positive cultures will usually give better results if incubated aerobically with added carbon dioxide. A candle jar is satisfactory for this purpose. Individual tubes may be sealed to give anaerobic + $CO_2$ or aerobic + $CO_2$ conditions by following the directions given under oxygen-requirement tests.

## CATALASE TEST

Prepare 3.0% $H_2O_2$ in distilled water from 30% $H_2O_2$ or use the commercial 3.0% solution. Store in refrigerator. The catalase test can be done by adding the $H_2O_2$ to: 1) a loopful of the organism from a broth culture placed on a glass slide, 2) colonies on a BHIA or other non-blood-containing medium plate or 3) directly to the growth on a BHIA slant. The latter methods are preferable to the use of a broth culture, and the slants probably give the most reproducible results. If catalase is present, there will be an *ACTIVE* production of bubbles. A positive control should always be included. All anaerobic cultures should be held aerobically for 30 minutes before testing.

## O-F TEST

*Medium:* O-F Basal medium with 1.0% glucose.
*Procedure:* Inoculate two tubes of medium using a capillary pipette. Stab to the base of the medium and gently withdraw the pipette while expelling a drop of inoculum. Cover one tube of the pair with 4-5 mm sterile petrolatum. Incubate and observe for acid production at 3, 7 and 14 days. Compare with an uninoculated tube.

## INDOLE TEST

*Medium:* Indole-Nitrite Broth.
*Procedure:* Inoculate two tubes. Test at 3 and 7 days for indole by using the procedure of adding xylol and Ehrlich's reagent.

## NITRATE REDUCTION

*Medium:* Indole-Nitrite Broth.
*Procedure:* Inoculate two tubes. Test at 3 and 7 days for nitrate reduction. Add 5 drops of reagent 1 and then 5 drops of reagent 2. The development of a brick-red color shows that nitrite is present. If no color develops, the organism has not reduced nitrate or has reduced both nitrate and nitrite. Add a small amount of zinc powder. If a red color develops, nitrate is still present, and the test is negative (nitrate was not reduced). If the tube remains colorless the test is positive (both nitrate and nitrite reduced).
Reagent 1. 8.0 g sulfanilic acid + 1000 ml 5 N acetic acid
Reagent 2. 5.0 g $\alpha$-naphthylamine + 1000 ml 5 N acetic acid

## NITRITE REDUCTION

*Medium:* Trypticase Soy Broth + 0.01% $KNO_2$. Most tests have been done with 0.01% $KNO_2$, but recent experience suggests that this may be toxic to some organisms and that better results are obtained with 0.001% $KNO_2$.
*Procedure:* Incubate for 7 days and test with standard nitrate reagents as described under nitrate reduction. If a red color develops, nitrite is not reduced (test negative). If no color develops, nitrite is reduced (test positive).

## ESCULIN HYDROLYSIS

*Medium:*

| | | |
|---|---|---|
| Heart Infusion Broth | 25.0 g | |
| Esculin | 1.0 g | |
| Agar | 1.0 g | |
| Distilled water | 1000.0 ml pH 7.0 | |

*Procedure:* Inoculate by stabbing with a capillary pipette. Place tip of pipette at the bottom of the tube and withdraw slowly while expelling a drop of inoculum. Incubate for 7 days. Add a few drops of 1.0% ferric citrate to the top of the tube. Development of a brownish black color indicates hydrolysis.

## STARCH HYDROLYSIS

*Medium:* Nutrient agar + 0.5% soluble or insoluble starch. To prepare, suspend the starch in 40 ml cold water and add to the melted nutrient medium. Autoclave and pour into petri plates.

*Procedure:* Using a capillary pipette, transfer 1 drop of inoculum to one side of the plate and pull the drop across the surface of the medium. Two organisms may be placed on a plate. Inoculate three plates for each culture. Test at 24 hrs., 3 days and 7 days. To test, flood the plate with 2-4 ml Gram's iodine. No hydrolysis: entire plate turns blue. Complete hydrolysis: colorless area around colony or streak. Partial hydrolysis: pink area around colony or streak.

## GELATIN HYDROLYSIS

1. **Tube method.** *Medium:* Thiogel Medium (BBL).

*Procedure:* Boil medium for 2 minutes and cool just before inoculation. Inoculate by stabbing with a capillary pipette while still fluid. Allow to stand until solidified. Incubate tests and an uninoculated tube as a control and observe for liquefaction after 7 and 14 days. Place tests in the refrigerator and read for liquefaction as soon as the control tube has hardened.

2. **Plate method.** *Medium:* BHI broth + 15 g agar and 4 g gelatin per liter. Soak gelatin in 40 ml cold water and mix with the melted agar. Dispense in petri plates.

*Procedure:* Using a capillary pipette, draw 1 drop of inoculum across the plate in a straight line. Two organisms may be tested on a plate. Inoculate two plates for each culture. Test for hydrolysis after 7 and 14 days by flooding the plate with 2-4 ml $HgCl_2$ (15 g $HgCl_2$ + 20 ml conc HCl + 100 ml distilled water). The medium will become cloudy. Where hydrolysis has occurred, there will be a clear zone around the colony or streak.

## UREASE PRODUCTION

*Medium:*
1. Christensen's Urea Agar.
2. Thioglycollate Urease Broth.
   a. Fluid Thioglycollate Medium without dextrose or indicator + 2 g yeast extract per liter. Tube in 8.0 ml amounts and autoclave.
   b. Urea Broth. Add 31 g to 100 ml distilled water: filter sterilize.
   c. For the complete medium, add 1.0 ml urea broth to each 8.0 ml of thioglycollate broth.

*Procedure:* Inoculate the agar by flooding the slant with 1-2 drops of inoculum. Inoculate broth with 0.1 ml suspension. In both cases, observe for change in the phenol red indicator to a deep red indicating the production of an alkaline reaction due to the splitting of urea.

## HYDROGEN SULFIDE PRODUCTION

*Medium:*
1. BHIA slants.
2. Triple sugar iron agar slants.

*Procedure:* Inoculate slant with a drop of inoculum suspension from a capillary pipette. Suspend a lead acetate paper strip over the slant. Read for blackening of the paper after 7 and 14 days.

## HIPPURATE HYDROLYSIS

*Medium:* TSB + 1.0% sodium hippurate.
*Procedure:* Inoculate with 0.1 ml and incubate for 7 days. Add 1.0 ml of the culture supernate to 1.5 ml 50% $H_2SO_4$. Let stand for 4 hrs. and look for flat crystals of benzoic acid.

## CASEIN HYDROLYSIS

*Medium:* TSA + Sterile skim milk. Sufficient milk is added to produce a barely turbid medium (approx. 10 ml/500 ml medium). The milk is added to the melted and cooled medium just before pouring the plates.
*Procedure:* Streak plates and incubate for 7 days. Observe for clear zones around growth.

## DNase PRODUCTION

*Medium:* DNase Test agar + 0.2% glucose.
*Procedure:* Streak plates and incubate for 7 days. Flood plate with normal HCl and observe for clear zone around colonies.

## MILK

*Medium:* Place a few iron filings in a tube and add 8.0 ml milk. Use dehydrated skim milk + 5 g yeast extract per liter or whole nonhomogenized milk.
*Procedure:* Incubate for 14 days and read for acid (A), coagulation (C), gas (G), digestion (D) and blackening (B). Test for acidity in a spot plate with brom cresol purple indicator or add a drop of 0.1% brom cresol purple to the tubes.

## FERMENTATION TESTS

*Medium:*  1. Thioglycollate Fermentation Base (TFB). Use for *Actinomyces* and *Arachnia.*

| | |
|---|---|
| Fluid Thioglycollate Medium without dextrose or indicator | 1 liter |
| Yeast extract (glucose free) | 2.0 g |
| Brom cresol purple (1.0% aqueous) | 2.0 ml |

Tube in 9.0 ml amounts without gas trap vials.

2. Meat-Extract Peptone Base. Use for *Rothia.*

| | |
|---|---|
| Beef extract (Difco) | 3.0 g |
| Peptone (Bacto) | 10.0 g |
| NaCl | 5.0 g |
| Dist. $H_2O$ | 1000 ml |
| Andrade's indicator pH 7.4 | 10.0 ml |

Tube in 4.5 ml amounts without gas trap vials.

All carbohydrates are added to the medium aseptically to give a final concentration of 1.0%, except soluble starch, dextrin, glycogen, inulin, inositol and salicin, which are used at 0.5%.

**Carbohydrates.** The compounds given in list 1 are those suggested for speciation of *Actinomyces* and *Arachnia* (for *Rothia* see Table 13-3) in chapter 13. List 2 contains the remainder of the compounds used by the Subgroup. In this list compounds used routinely are italicized.

1. *Speciation*

Adonitol
Arabinose
Dulcitol
Glucose
Glycerol
Mannitol
Raffinose
Salicin
Starch, soluble
Sucrose
Xylose

2. *Additional*

*Cellobiose*
Dextrin
Erythritol
Fructose
Galactose
Glycogen
α-methyl-d-glucoside
*Inositol*
Inulin
*Lactose*

Maltose
*Mannose*
α-methyl-D-mannoside
Melezitose
Melibiose
*Rhamnose*
*Ribose*
Sorbitol
*Trehalose*

## END PRODUCTS OF GLUCOSE FERMENTATION

Chromatographic procedures for the determination of fermentation end products are described in detail in the VPI Anaerobe Laboratory Manual (201), Anaerobic Bacteriology Manual (447) and by Li and Georg (284). The procedures described for anaerobes are applicable to the actinomycetes.

## FLUORESCENT ANTIBODY TECHNIQUES

**Preparation Procedures.** (See Serology section for immunization procedures; FITC conjugated antisera for *A. israelii* and *A. naeslundii* are available from the Biological Reagent Section, CDC, Atlanta, Ga.)
1. *Clinical material.* Make two smears of the material on a clean glass slide containing two marked circles. Air dry and fix by flooding the slide with methanol for 1 minute. Pour off alcohol and allow to air dry.
2. *Cultures.* Make smears of a suspension of the organism obtained by centrifuging a broth culture or from a plate culture. If thioglycollate cultures are used, remove a small portion of the growth with a capillary pipette and resuspend in a small amount of sterile saline. Air dry smears and gently heat fix.
   **Staining Procedures.**
1. Place 1 drop of conjugated antiserum on each fixed smear. Incubate under a moist chamber for 30 minutes at room temperature.
2. Remove excess conjugate by tilting slide onto a paper towel. Wash in two changes of pH 7.2 buffer (FTA Haemagglutination Buffer—BBL) for 5 minutes each.
3. Counterstain in 0.5% aqueous Evans Blue for 5 minutes. Rhodamine labeled bovine albumin can also be used as a counterstain by following directions of the manufacturer.
4. Dip briefly in distilled water to remove excess Evans Blue and wash in two changes of pH 9.0 buffer for 1 minute each.
5. Air dry or blot carefully (not recommended for clinical material).
6. Place 1 drop of buffered glycerol mounting fluid (9 parts c.p. glycerol + 1 part Ph 9.0 buffer) on smear and place a #1 coverslip over it.
7. Examine under a microscope equipped for FA work. We have found that a 54X oil immersion objective is most useful in FA studies.

## MAINTENANCE OF CULTURES

For cultures in routine use, weekly transfer in Thioglycollate Broth containing 0.2% rabbit serum has proven satisfactory. For long-term storage, cultures may be frozen or lyophilized.

**1. Freezing.** Place 3-5 ml BHI broth in small, screw-capped test tubes or vials: autoclave. Inoculate with 0.1 ml broth culture. Incubate for 3-5 days. As soon as there is good growth in the tubes, tighten caps and place in freezer at −70°C. Frozen cultures have remained viable for 1 year.

**2. Lyophilization.** Use BHI broth cultures (3-5 days) with good growth, centrifuge and re-suspend in sterile skim milk. Usually a 10.0 ml broth culture yields a heavy suspension in 0.5 ml of milk. Lyophilize by standard techniques. We have had viable cultures after 10 years of storage at room temperature.

# 15

# Bibliography

1. Abraham, E. P., E. Chain, C. M. Fletcher, A. D. Gardiner, N. G. Heatley, M. A. Jennings, and H. W. Florey. 1941. Further observations on penicillin. Lancet. *241:* 177-188.
2. Adams, A. R., J. M. Jackson, J. Scopa, C. K. Lane, and C. R. Wilson. 1971. Nocardiosis. Diagnosis and management with a report of three cases. Med. J. Aust. *1:* 669-674.
3. Adar, R., E. Antebi, R. David, and M. Mozes. 1972. Abdominal actinomycosis. Isr. J. Med. Sci. *8:* 148-153.
4. Affronti, L. F. 1959. Purified protein derivatives and other antigens prepared from atypical acid-fast bacilli and *Nocardia asteroides*. Amer. Rev. Tuberc. *79:* 284-295.
5. Afrikian, E. G., S. S. Julian, and L. A. Bulla. 1973. Scanning electron microscopy of bacterial colonies. Appl. Microbiol. *26:* 934-937.
6. Al-Doory, Y. 1965. Fluorescent antibody studies with *Nocardia asteroides*. I. Preliminary attempts in removing cross-reacting factors. Sabouraudia. *4:* 135-142.
7. Altman, N. H., and J. D. Small. 1973. Actinomycosis in a primate confirmed by fluorescent antibody technics in formalin fixed tissue. Lab. Anim. Sci. *23:* 696-700.
8. Anhalt, M., and R. Scott. 1970. Primary unilateral renal actinomycosis: Case report. J. Urol. *103:* 126-129.
9. Anonymous. 1963-1973. Reports from members of the Subgroup on Taxonomy of Micro-aerophilic Actinomycetes Presented at ASM meetings and International Congresses in Moscow and Mexico City.
10. Baboolal, R. 1969. Cell wall analyses of oral filamentous bacteria. J. Gen. Microbiol. *58:* 217-226.
11. Baboolal, R. 1972. A study of the enzyme patterns of some oral filamentous bacteria by starch gel electrophoresis. Arch. Oral Biol. *17:* 691-700.
12. Bach, M. C., L. D. Sabath, and M. Finland. 1973. Susceptibility of *Nocardia asteroides* to 45 antimicrobial agents *in vitro*. Antimicrob. Ag. Chemother. *3:* 1-8.
13. Baird-Parker, A. C., and G. H. G. Davis. 1958. The morphology of *Leptotrichia* species. J. Gen. Microbiol. *19:* 446-450.
14. Baker, R. D. 1971. Human Infection with Fungi, Actinomycetes and Algae. Springer-Verlag, New York.
15. Bates, M., and G. Cruickshank. 1957. Thoracic actinomycosis. Thorax. *12:* 99-124.
16. Batshon, B. A., O. C. Brosius, and J. C. Snyder. 1971. A case report of *Nocardia asteroides* of the eye. Mycologia. *63:* 459-461.
17. Batty, I. 1958. *Actinomyces odontolyticus*, a new species of actinomycete regularly isolated from deep carious dentine. J. Pathol. Bacteriol. *75:* 455-459.

18. Baylin, G. L., and J. M. Wear. 1953. Blastomycosis and actinomycosis of the spine. Amer. J. Roentgenol. Radium Ther. Nucl. Med. *69:* 395-398.
19. Bayne-Jones, S. 1925. Club-formation by *Actinomyces hominis* in glucose broth, with a note on *B. actinomycetem-comitans.* J. Bacteriol. *10:* 569-576.
20. Becker, B., M. P. Lechevalier, R. E. Gordon, and H. A. Lechevalier. 1964. Rapid differentiation between *Nocardia* and *Streptomyces* by paper chromatography of whole-cell hydrolysates. Appl. Microbiol. *12:* 421-423.
21. Berd, D. 1973. *Nocardia asteroides.* A taxonomic study with clinical correlation. Amer. Rev. Resp. Dis. *108:* 909-917.
22. Berd, D. 1973. *Nocardia brasiliensis* infection in the United States. Am. J. Clin. Pathol. *60:* 254-258.
23. Berd, D. 1973. Laboratory identification of clinically important aerobic actinomycetes. Appl. Microbiol. *25:* 665-681.
24. Bergey, D. H. 1907. Aktinomyces der mundhole. Zentralbl. Bakteriol. Parastenk. Infektionskr. Abt. I. Ref. *40:* 361.
25. Bevan, A. D. 1905. Treatment of actinomycosis and blastomycosis with copper salts. J. Amer. Med. Ass. *45:* 1492-1493.
26. Bevis, M. L. 1961. Serological studies on *Mycobacterium, Nocardia, Actinomyces* and *fungi.* Sabouraudia. *1:* 154-160.
27. Bier, L. C., and W. C. de Araujo. 1971. Isolamento e caracterizacão de *Actinomyces viscosus* do sulco gengival humano. Rev. Microbiol. *2:* 3-15.
28. Biever, L. J. 1967. A bacteriologic study of abscesses of swine and cattle. M. S. Thesis. South Dakota State University.
29. Biever, L. J., G. W. Robertstad, K. Van Steenbergh, E. E. Scheetz, and G. F. Kennedy. 1969. Actinomycosis in a bovine lung. Amer. J. Vet. Res. *30:* 1063-1066.
30. Black, W. A., and D. A. McNellis. 1970. Sensitivity of *Nocardia* to trimethoprin and sulphonamides *in vitro.* J. Clin. Pathol. *23:* 423-426.
31. Black, W. A., and D. A. McNellis. 1971. Comparative *in vitro* sensitivity of *Nocardia* species to fusidic acid and sulphonamides. J. Med. Microbiol. *4:* 293-295.
32. Blanchard, R. 1896. Parasites végétaux à l'exclusion des bacteries. Traite de Pathologie General. G. Mason, Paris. *2:* 811-926.
33. Blank, C. H., and L. K. Georg. 1968. The use of fluorescent antibody methods for the detection and identification of *Actinomyces* species in clinical material. J. Lab. Clin. Med. *71:* 283-293.
34. Blevins, A., C. Semolic, M. Sukany, and D. Armstrong. 1974. Common isolation of *Rothia dentocariosa* from clinical specimens studied in the microbiology laboratory. Abst. ASM. *73:* 117.
35. Bojalil, L. F., and J. Cerbon. 1959. Schema for the differentiation of *Nocardia asteroides* and *Nocardia brasiliensis.* J. Bacteriol. *78:* 852-857.
36. Bojalil, L. F., and M. Magnusson. 1963. Specificity of skin reactions of humans to *Nocardia* sensitins. Amer. Rev. Resp. Dis. *88:* 409-411.
37. Bollinger, O. 1877. Ueber eine neu pilzkrankheit beim rinde. Zentralbl. Med. Wiss. *15:* 481-485.
38. Bolton, C. F., and E. M. Ashenhurst. 1964. Actinomycosis of the brain: Case report and review of literature. Can. Med. Ass. J. *90:* 922-928.
39. Bonner, M., and M. Bonner. 1971. *Actinomyces.* John Wright and Sons, Bristol, 2 ed.
40. Boone, C. J., and L. Pine. 1968. Rapid method for characterization of actinomycetes by cell wall composition. Appl. Microbiol. *16:* 279-284.
41. Bostroem, E. 1891. Untersuchungen über die aktinomykose des menschen. Beitr. Pathol. Anat. Allg. Pathol. *9:* 1-240.
42. Bowden, G. H., and J. M. Hardie. 1973. Commensal and pathogenic *Actinomyces* species in man. *In Actinomycetales:* Characteristics and practical importance, ed. G. Sykes and F. A. Skinner. Academic Press, New York.
43. Boyd, A., and R. A. D. Williams. 1971. Estimation of the volumes of bacterial cells by scanning electron microscopy. Arch. Oral Biol. *16:* 259-267.

44. Boyd, M. F., and E. D. Crutchfield. 1921. A contribution to the study of mycetoma in North America. Amer. J. Trop. Med. *1:* 215-289.

45. Bradley, P. 1971. Actinomycosis of the temporomandibular joint. Brit. J. Oral Surg. *9:* 54-56.

46. Bragg, S., L. Georg, and A. Ibrahim. 1972. Determination of a new serotype of *Actinomyces naeslundii.* Abst. ASM *72:* 38.

47. Breed, R. S., and H. J. Conn. 1919. The nomenclature of the *Actinomycetaceae.* J. Bacteriol. *4:* 585-602.

48. Brock, D. W., and L. K. Georg. 1969. Characterization of *Actinomyces israelii.* Serotypes 1 and 2. J. Bacteriol. *97:* 589-593.

49. Brock, D. W., and L. K. Georg. 1969. Determination and analysis of *Actinomyces israelii* serotypes by fluorescent-antibody procedures. J. Bacteriol. *97:* 581-588.

50. Brock, D. W., L. K. Georg, J. M. Brown, and M. D. Hicklin. 1973. Actinomycosis caused by *Arachnia propionica.* Amer. J. Clin. Pathol. *59:* 66-77.

51. Brodie, B., and J. P. Manning. 1967. Isoniazed treatment in actinomycosis. Med. Vet. Pract. (Noll). *48:* 70-71.

52. Brown, J. M., L. K. Georg, and L. C. Waters. 1969. Laboratory identification of *Rothia dentocariosa* and its occurrence in human clinical material. Appl. Microbiol. *17:* 150-156.

53. Brown, J. R., and F. von Lichtenberg. 1970. Experimental actinomycosis in mice. Arch. Pathol. *90:* 391-402.

54. Brown, O., and J. B. Clark. 1966. Evidence for the pentose cycle in *Nocardia corallina.* Proc. Soc. Exp. Biol. Med. *122:* 887-890.

55. Brownell, G. H., J. N. Adams, and S. G. Bradley. 1967. Growth and characteristics of nocardiophages for *Nocardia canicruria* and *Nocardia erythropolis* mating types. J. Gen. Microbiol. *47:* 247-256.

56. Buchanan, B. B., and L. Pine. 1962. Characterization of a propionic acid producing actinomycete, *Actinomyces propionicus,* sp. nov. J. Gen. Microbiol. *28:* 305-323.

57. Buchanan, B. B., and L. Pine. 1963. Factors influencing the fermentation and growth of an atypical strain of *Actinomyces naeslundii.* Sabouraudia *3:* 26-39.

58. Buchanan, B. B., and L. Pine. 1965. Relationship of carbon dioxide to aspartic acid and glutamic acid in *Actinomyces naeslundii.* J. Bacteriol. *89:* 729-733.

59. Buchanan, B. B., and L. Pine. 1967. Path of glucose breakdown and cell yields of a facultative anaerobe, *Actinomyces naeslundii.* J. Gen. Microbiol. *46:* 225-236.

60. Buchanan, R. E. 1918. Studies in the classification and nomenclature of the bacteria. VIII. The subgroups and genera of the *Actinomycetales.* J. Bacteriol. *3:* 403-406.

61. Buchanan, R. E., J. G. Holt, and E. F. Lessel. 1966. Index Bergeyana. Williams and Wilkins Co., Baltimore.

62. Buchs, H. 1963. Zur klinik und therapie der cervicofaciaten aktinomykose. Deut. Zahnaerztl. Z. *18:* 1060-1075.

63. Bujwid, O. 1889. Ueber die reinkultur des actinomyces. Zentralbl. Bakteriol. Parasitenk. Infektionskr. *6:* 630-633.

64. Burdon, D. W. 1971. Nocardiosis after appendectomy. Brit. Med. J. *1:* 538.

65. Burnett, G. W., and H. W. Scherp. 1968. Oral Microbiology and Infectious Disease. Williams and Wilkins Co., Baltimore.

66. Caldwell, J. L. 1971. Actinomycosis treated with cephalothin. South. Med. J. *64:* 987 and 991.

67. Calmes, R., and S. J. Deal. 1973. Fatty acid activation by a lipophilic bacterium. J. Bacteriol. *114:* 249-256.

68. Castellani, A., and A. J. Chalmers. 1913. Manual of Tropical Medicine. 2nd ed. Balliere, Tindall and Cox, London, 1-1719.

69. Causey, W. A., and B. Sieger. 1973. Systemic nocardiosis caused by *Nocardia brasiliensis.* Am. Rev. Resp. Dis. *109:* 134-137.

70. Christie, A. O., and J. W. Porteus. 1962. The cultivation of a single strain of *Actinomyces israelii* in a simplified and chemically defined medium. J. Gen. Microbiol. *28:* 443-454.

71. Christie, A. O., and J. W. Porteus. 1962. The growth factor requirements of the Wills strain

of *Actinomyces israelii* growing in a chemically defined medium. J. Gen. Microbiol. *28:* 455-460.

72. Christie, A. O., and J. W. Porteus. 1962. Growth of several strains of *Actinomyces israelii* in a chemically defined medium. Nature. *195:* 408-409.

73. Cock, D. J., and W. H. Bowen. 1967. Occurrence of *Bacterionema matruchotii* and *Bacteroides melaninogenicus* in gingival plaque from monkeys. J. Periodontal Res. *2:* 36-39.

74. Cohen, M. L., E. B. Weiss, and A. P. Monaco. 1971. Successful treatment of *Pneumocystis carinii* and *Nocardia asteroides* in a renal transplant patient. Amer. J. Med. *50:* 269-276.

75. Cohn, F. 1875. Untersuchungen über Bacterien II. Beitrage zur Biologie der Pflanzen *III:* 141-207.

76. Colebrook, L. 1921. A report on 25 cases of actinomycosis with special reference to vaccine therapy. Lancet. *1:* 893-899.

77. Coleman, R. M., and L. K. Georg. 1969. Comparative pathogenicity of *Actinomyces naeslundii* and *Actinomyces israelii.* Appl. Microbiol. *18:* 427-432.

78. Coleman, R. M., L. K. Georg, and A. R. Rozzell. 1969. *Actinomyces naeslundii* as an agent of human actinomycosis. Appl. Microbiol. *18:* 420-426.

79. Collins, P. A., M. A. Gerencser, and J. M. Slack. 1973. Enumeration and identification of *Actinomycetaceae* in human dental calculus using the fluorescent antibody technique. Arch. Oral Biol. *18:* 145-153.

80. Coodley, E. L. 1969. Actinomycosis: Clinical diagnosis and management. Postgrad. Med. *46:* 73-78.

81. Cope, V. Z. 1915. A clinical study of actinomycosis with illustrative cases. Brit. J. Surg. *3:* 55-81.

82. Cope, V. Z. 1938. Actinomycosis. Oxford University Press, London.

83. Cope, V. Z. 1951. Actinomycosis of bone with special reference to infection of vertebral column. J. Bone Joint Surg. *33-B:* 205-214.

84. Corda, A. C. J. 1839. Pracht-Flora Europaeischer Schimmelbildungen. Gerhard Fleischer, Leipzig.

85. Costigan, P. G. 1947. A case of actinomycosis treated with streptomycin. Can. Med. Ass. J. *56:* 431.

86. Coykendall, R. L., T. W. Lee, and A. T. Brown. 1974. Isolation and base content of deoxyribonucleic acid (DNA) from *Actinomyces viscosus* strains. Abst. IADR. *74:* 73.

87. Crawford, J. J. 1971. Interaction of *Actinomyces* organisms with cationic polypeptides. I. Histochemical studies of infected human and animal tissues. Infect. Immunity. *4:* 632-641.

88. Cullen, C. H., and M. E. Sharp. 1951. Infection of wounds with *Actinomyces.* J. Bone Joint Surg. *33-B:* 221-227.

89. Cummins, C. S. 1962. Chemical composition and antigenic structure of cell walls of *Corynebacterium, Mycobacterium, Nocardia, Actinomyces* and *Arthrobacter.* J. Gen. Microbiol. *28:* 35-50.

90. Cummins, C. S. 1965. Ornithine in mucopeptide of gram-positive cell walls. Nature. *206:* 1272.

91. Cummins, C. S., and H. Harris. 1958. Studies on the cell wall composition and taxonomy of *Actinomycetales* and related groups. J. Gen. Microbiol. *18:* 173-189.

92. Cummins, C. S., and H. Harris. 1959. Cell-wall composition in strains of *Actinomyces* isolated from human and bovine lesions. J. Gen. Microbiol. *21:* ii.

93. Davis, G. H. G., and A. C. Baird-Parker. 1959. *Leptotrichia buccalis.* Brit. Dent. J. *106:* 70-73.

94. Davis, G. H. G., and A. C. Baird-Parker. 1959. The classification of certain filamentous bacteria with respect to their chemical composition. J. Gen. Microbiol. *21:* 612-621.

95. Davis, G. H. G., and J. H. Freer. 1960. Studies upon an oral aerobic actinomycete. J. Gen. Microbiol. *23:* 163-178.

96. Davis, J. W., L. H. Karstad, and V. O. Trainer. 1970. Infectious Diseases of Wild Animals. Iowa State University Press, Ames.

97. DeWeese, M. S., M. A. Gerencser, and J. M. Slack. 1968. Quantitative analysis of *Actinomyces* cell walls. Appl. Microbiol. *16:* 1713-1718.

98. Dobell, C. 1932. Antony van Leeuwenhoek and His "Little Animals." Staples Press Limited, London.

99. Dorph-Petersen, L., and J. J. Pindborg. 1954. Actinomycosis of the tongue. Oral Surg. Oral Med. Oral Pathol. *7:* 1178-1182.

100. Drake, C. H., and A. T. Henrici. 1943. *Nocardia asteroides.* Its pathogenicity and allergic properties. Amer. Rev. Tuberc. *48:* 184-198.

101. Duda, J. J., and J. M. Slack. 1972. Ultrastructural studies on the genus *Actinomyces.* J. Gen. Microbiol. *71:* 63-68.

102. Duff, R. B., and D. M. Webley. 1959. Production of pentose intermediates during growth of *Nocardia opaca* and other saprophytic soil nocardias and mycobacteria. Biochim. Biophys. Acta. *34:* 398-406.

103. Duncan, J. A. 1965. Abdominal actinomycosis: Changed concepts? Amer. J. Surg. *110:* 148-152.

104. Durie, E. B. 1958. A critical survey of mycological research and literature for the years 1946-1956 in Australia. Mycopathol. Mycol. Appl. *9:* 80-96.

105. Dutton, W. P., and A. P. Inclan. 1968. Cardiac actinomycosis. Dis. Chest. *54:* 463-465.

106. Dyson, J. E., and J. M. Slack. 1963. Improved antigens for skin testing in nocardiosis. I. Alcohol precipitates of culture supernates. Amer. Rev. Resp. Dis. *88:* 80-86.

107. Eastridge, C. E., J. R. Prather, F. A. Hughes, J. M. Young, and J. J. McCaughan. 1972. Actinomycosis: A 24 year experience. South. Med. J. *65:* 839-843.

108. Edwards, M. R., and M. A. Gordon. 1962. Membrane systems of *Actinomyces bovis.* Intern. Conf. Electron Microscopy, 5th. *2:* UU-3.

109. Ellis, P. P., S. C. Bausor, and J. M. Fulmer. 1961. Streptothrix canaliculitis. Amer. J. Ophthalmol. *52:* 36-43.

110. Emmons, C. W. 1938. The isolation of *Actinomyces bovis* from tonsillar granules. Publ. Health Rep. *53:* 1967-1975.

111. Emmons, C. W., C. H. Binford, and J. P. Utz. 1970. Medical Mycology. 2nd ed. Lea and Febiger, Philadelphia.

112. Ennever, J. 1963. Microbiologic calcification. Ann. N.Y. Acad. Sci. *109:* 4-13.

113. Eppinger, H. 1891. Uber eine neue pathogene *Cladothrix* und eine durch sie hervorgerufene pseudotuberculosis (cladothrichica). Beitr. Pathol. Anat. Allg. Pathol. *9:* 287-328.

114. Erikson, D. 1935. The pathogenic aerobic organisms of the *Actinomyces* group. Med. Res. Coun. Spec. Rep. Ser. No. *203:* 5-61.

115. Erikson, D. 1940. Pathogenic anaerobic organisms of the *Actinomyces* group. Med. Res. Council Spec. Rep. Ser. No. *240:* 1-63.

116. Ernst, J., and E. Ratjen. 1971. Actinomycosis of the spine. Acta Orthop. Scand. *42:* 35-44.

117. Erving, W. G. 1902. *Actinomycosis hominis* in America, with report of six cases. Bull. Johns Hopkins Hosp. *13:* 261-268.

118. Everts, E. C. 1970. Cervicofacial actinomycosis. Arch. Otolaryngol. *92:* 468-474.

119. Falk, J. 1969. Rough thylakoids: Polysomes attached to chloroplast membranes. J. Cell. Biol. *42:* 582-587.

120. Farris, E. M., and R. V. Douglas. 1947. Abdominal actinomycosis. Arch. Surg. *54:* 434-444.

121. Fawi, M. F. 1964. Complement fixing antibodies in nocardiosis with special reference to dogs. Sabouraudia. *3:* 303-305.

122. Fetter, B. F., G. K. Klintworth, and W. S. Hendry. 1967. Mycoses of the Central Nervous System. Williams and Wilkins Co., Baltimore.

123. Fletcher, R. 1956. A rare case of chronic otorrhea with intracranial complications. Laryngoscope. *66:* 702-710.

124. Florey, M. E., and H. W. Florey. 1943. General and local administration of penicillin. Lancet. *244:* 387-397.

125. Flynn, M. W., and B. Felson. 1970. The roentgen manifestations of thoracic actinomycosis. Amer. J. Roentgenol. Radium Ther. Nucl. Med. *110:* 707-716.

126. Foley, T. F., D. E. Dines, and C. T. Dolan. 1971. Pulmonary actinomycosis. Minn. Med. *54:* 593-598.

127. Frank, R. M., B. Guillo, and H. Llory. 1972. Caries dentaries chez le rat gnotobiote inoculé avec *Actinomyces viscosus* et *Actinomyces naeslundii.* Arch. Oral Biol. *17:* 1249-1253.

128. Franke, F. 1973. The etiology of actinomycosis of the mammary gland of the pig. Zentralbl. Bakteriol. Parasitenk. Infektionskr. Hyg. Abt. I. Orig. *223:* 111-124.

129. Frazier, P. D., and B. O. Fowler. 1967. X-ray diffraction, and infrared study of the "sulphur granules" of *Actinomyces bovis.* J. Gen. Microbiol. *46:* 445-450.

130. Freer, J., K. S. Kim, M. R. Krauss, L. Beaman, and K. Barksdale. 1969. Ultrastructural changes in bacteria isolated from cases of leprosy. J. Bacteriol. *100:* 1062-1075.

131. Fry, G. A., M. J. Martin, W. H. Dearing, and C. E. Culp. 1965. Primary actinomycosis of the rectum with multiple perianal and perineal fistulae. Proc. Staff Meet. Mayo Clin. *40:* 296-299.

132. Gale, D., and C. A. Waldron. 1955. Experimental actinomycosis with *Actinomyces israelii.* J. Infect. Dis. *97:* 251-261.

133. Gardiner, S. S. 1936. Actinomycosis of the head and neck. Aust. N. Z. J. Surg. *6:* 186-224.

134. Gardiner, S. S. 1943. Actinomycosis of the urinary system. Aust. N. Z. J. Surg. *12:* 210-284.

135. Geister, R. S., and E. Meyer. 1951. The effect of aureomycin and penicillin on experimental actinomycosis infections in mice. J. Lab. Clin. Med. *38:* 101-111.

136. Georg, L. K. 1970. Diagnostic procedures for the isolation and identification of the etiologic agents of actinomycosis. Proc. Int. Symp. on Mycosis. WHO Washington, D.C.

137. Georg, L. K. 1974. Genus *Rothia,* Bergey's Manual of Determinative Bacteriology. 8th ed. N. E. Gibbons, ed. Williams and Wilkins Co., Baltimore.

138. Georg, L. K., L. Ajello, C. McDurmont, and T. S. Hosty. 1961. The identification of *Nocardia asteroides* and *Nocardia brasiliensis.* Amer. Rev. Resp. Dis. *84:* 337-347.

139. Georg, L. K., and J. M. Brown. 1967. *Rothia,* gen. nov. An aerobic genus of the family *Actinomycetaceae.* Int. J. Syst. Bacteriol. *17:* 79-88.

140. Georg, L. K., J. M. Brown, H. J. Baker, and G. H. Cassell. 1972. *Actinomyces viscosus* as an agent of actinomycosis in the dog. Amer. J. Vet. Res. *33:* 1457-1470.

141. Georg, L. K., and R. M. Coleman. 1970. Comparative pathogenicity of various *Actinomyces* species. The *Actinomycetales.* Jena Inter. Symp. Taxon. *1:* 35-45.

142. Georg, L. K., L. Pine, and M. A. Gerencser. 1969. *Actinomyces viscosus* comb. nov. A catalase positive, facultative member of the genus *Actinomyces.* Int. J. Syst. Bacteriol. *19:* 291-293.

143. Georg, L. K., G. W. Robertstad, and S. A. Brinkman. 1964. Identification of species of *Actinomyces.* J. Bacteriol. *88:* 477-490.

144. Georg, L. K., G. W. Robertstad, S. A. Brinkman, and M. D. Hicklin. 1965. A new pathogenic anaerobic *Actinomyces* species. J. Infect. Dis. *115:* 88-99.

145. Gerencser, M. A., and J. M. Slack. 1967. Isolation and characterization of *Actinomyces propionicus.* J. Bacteriol. *94:* 109-115.

146. Gerencser, M. A., and J. M. Slack. 1969. Identification of human strains of *Actinomyces viscosus.* Appl. Microbiol. *18:* 80-87.

147. Gerszten, E., M. J. Allison, and H. P. Dalton. 1969. A ten-year study of mycotic infections in a Virginia general hospital. Amer. J. Clin. Pathol. *52:* 445-450.

148. Gibbons, R. J. 1972. Microbiol. ecological models and dental diseases. Virginia Dent. J. *49:* 72-88.

149. Gibbons, R. J., K. S. Berman, P. Knoettner, and B. Kapsimilis. 1966. Dental caries and alveolar bone loss in gnotobiotic rats infected with capsule forming streptococci of human origin. Arch. Oral Biol. *11:* 549-560.

150. Gibbons, R. J., P. F. Depaola, D. M. Spinell, and Z. Skobe. 1974. Interdental localization of *Streptococcus mutans* as related to dental caries experience. Infect. Immunity. *9:* 481-488.

151. Gibbons, R. J., and M. Nygaard. 1970. Interbacterial aggregation of plaque bacteria. Arch. Oral Biol. *15:* 1397-1400.

152. Gibbons, R. J., and D. M. Spinell. 1969. Salivary-induced aggregation of plaque bacteria. *In* Dental Plaque, ed. W. G. McHugh. Livingstone, Edinburgh, Scotland.

153. Gibbons, R. J., and J. Van Houte. 1973. On the formation of dental plaque. J. Periodontol. *44:* 347-360.

154. Gilmour, M. N. 1974. Genus *Bacterionema*. Bergey's Manual of Determinative Bacteriology. 8th ed. N. E. Gibbons, ed. Williams and Wilkins Co., Baltimore.

155. Gilmour, M. N., and P. H. Beck. 1961. The classification of organisms termed *Leptotrichia* (Leptothrix) *buccalis*. II. Growth and biochemical characteristics of *Bacterionema matruchotii*. Bacteriol. Rev. *25:* 142-151.

156. Gilmour, M. N., and B. G. Bibby. 1966. A synthetic medium for *Bacterionema matruchotii*. J. Dent. Res. *45:* 158.

157. Gilmour, M. N., A. Howell, Jr., and B. G. Bibby. 1961. The classification of organisms termed *Leptotrichia* (Leptothrix) *buccalis*. I. Review of the literature and proposed separation into *Leptotrichia buccalis* (Trevisan, 1879) and *Bacterionema* gen. nov.; *B. matruchotii* (Mendel, 1919) comb. nov. Bacteriol. Rev. *25:* 131-141.

158. Gilmour, M. N., and P. A. Hunter. 1958. Isolation of an oral filamentous microorganism. J. Bacteriol. *76:* 294-300.

159. Girard, A. E., and B. H. Jacius. 1974. Ultrastructure of *Actinomyces viscosus* and *Actinomyces naeslundii*. Arch. Oral Biol. *19:* 71-79.

160. Gledhill, W. E., and L. E. Casida, Jr. 1969. Predominant catalase-negative soil bacteria. II. Occurrence and characterization of *Actinomyces humiferus*. sp. N. Appl. Microbiol. *18:* 114-121.

161. Gold, L., and E. E. Doyne. 1952. Actinomycosis with osteomyelitis of the alveolar process. Oral Surg. Oral Med. Oral Pathol. *5:* 1056-1063.

162. Goldstein, B. H., J. J. Sciubba, and D. M. Laskin. 1972. Actinomycosis of the maxilla: Review of literature and report of case. J. Oral Surg. *30:* 362-366.

163. Gonzalez-Ochoa, A. 1962. Mycetomas caused by *Nocardia brasiliensis*. Lab. Invest. *11:* 1118-1123.

164. Gonzalez-Ochoa, A. 1973. Virulence of nocardiae. Canad. J. Microbiol. *19:* 901-904.

165. Goodfellow, M. 1971. Numerical taxonomy of some nocardioform bacteria. J. Gen. Microbiol. *69:* 33-80.

166. Goodman, J. S., and M. G. Koenig. 1970. *Nocardia* infections in a general hospital. Ann. N.Y. Acad. Sci. *174:* 552-567.

167. Gordon, M. A., and H. M. DuBose. 1951. Anorectal actinomycosis. Amer. J. Clin. Pathol. *21:* 460-463.

168. Gordon, R. E., and W. A. Hagan. 1936. A study of some acidfast actinomycetes from soil with special reference to pathogenicity for animals. J. Infect. Dis. *59:* 200-206.

169. Gordon, R. E., and J. M. Mihm. 1959. A comparison of *Nocardia asteroides* and *Nocardia brasiliensis*. J. Gen. Microbiol. *20:* 129-135.

170. Gordon, R. E., and J. M. Mihm. 1967. Identification of *Nocardia caviae*. (Erikson) nov. comb. Ann. N.Y. Acad. Sci. *98:* 628-636.

171. Gossling, J., and J. M. Slack. 1974. Predominant gram-positive bacteria in human feces: Numbers, variety and persistence. Infect. Immunity *9:* 719-729.

172. Graefe, V. A. 1854. Koncretionen im unteren thränenröhrchen durch pilzbildung. Arch. Ophthalmol. *1:* 284-288.

173. Grässer, R. 1962. Mikroaerophile actinomyceten aus gesaugeaktinomykosen des schweiness. Zentralbl. Bakteriol. Parasitenk. Infektionskr. Hyg. Abt. I. Orig. *184:* 478-492.

174. Grässer, R. 1963. Untersuchungen über fermentative und serologische eigenschaften mikroaerophiler actinomyceten. Zentralbl. Bakteriol. Parasitenk. Infektionskr. Hyg. Abt. I. Orig. *188:* 251-263.

175. Graybill, J. R., and B. D. Silverman. 1969. Sulfur granules, second thoughts. Arch. Intern. Med. *123:* 430-432.

176. Grobert, M. J., and A. J. Bischoff. 1962. Actinomycosis of the testicle. Case report. J. Urol. *87:* 567-572.

177. Grossman, C. B., D. G. Bragg, and D. Armstrong. 1970. Roentgen manifestations of pulmonary nocardiosis. Radiology. *96:* 325-330.

178. Guidry, D. J. 1971. Actinomycosis in Human Infection with Fungi, Actinomycetes and Algae. R. D. Baker, ed. Springer-Verlag, New York.

179. Haass, E. 1906. Beitrag zur kenntnis der aktinomyceten. Zentralbl. Bakteriol. Parasitenk. Infectionskr. Hyg. Abt. 1 Orig. *40:* 180-186.

180. Halseth, W. L., and M. R. Reich. 1969. Pulmonary actinomycosis treated by lung resection. Dis. Chest. *55:* 119-121.

181. Hammond, B. F. 1970. Isolation and serological characterization of a cell wall antigen of *Rothia dentocariosa.* J. Bacteriol. *103:* 634-640.

182. Hammond, B. F. 1972. Cell wall analyses of *Rothia dentocariosa.* Abst. IADR. *72:* 108.

183. Hammond, B. F. 1974. Personal Communication.

184. Hammond, B. F., and K. S. Peindl. 1973. Catalase determination in gram-positive rods by disc-gel electrophoresis. Abst. ASM, *73:* 36.

185. Hammond, B. F., C. F. Steel, and K. Peindl. 1973. Occurrence of 6-deoxytalase in cell walls of plaque actinomycetes. J. Dent. Res. *52:* 88.

186. Harbitz, F., and N. B. Grondahl. 1911. Actinomycosis in Norway: Studies on the etiology, mode of infection and treatment. Amer. J. Med. Sci. *142:* 386-395.

187. Harsha, W. M. 1904. Actinomycosis of the jaw. Ann. Surg. *39:* 459-460.

188. Hartley, J. H., and W. E. Schatten. 1973. Cervicofacial actinomycosis. Plast. Reconstr. Surg. *51:* 44-47.

189. Harvey, J. C., J. R. Cantrell, and A. M. Fisher. 1957. Actinomycosis: Its recognition and treatment. Ann. Intern. Med. *46:* 868-885.

190. Harz, C. O. 1877. (In Bollinger.) O. Ueber eine neue pilzkrankheit bein rinde. Zentralbl. Med. Wiss. *15:* 481-485.

191. Harz, C. O. 1879. *Actinomyces bovis* ein neuer schimmel in dem geweben des rindes. Jahrebericht der K. Central-Thieraznei Schule in München für 1877-1878. *5:* 125-140.

192. Hazen, E. L., and G. N. Little. 1958. *Actinomyces bovis* and "anaerobic diphtheroids" pathogenicity for hamsters and some other differentiating characteristics. J. Lab. Clin. Med. *51:* 968-976.

193. Hazen, E. L., G. L. Little, and H. Resnick. 1952. The hamster as a vehicle for the demonstration of pathogenicity of *Actinomyces bovis.* J. Lab. Clin. Med. *40:* 914-918.

194. Hebert, G. A., B. Pittman, R. M. McKinney, and W. B. Cherry. 1972. The preparation and physicochemical characterization of fluorescent antibody reagents. U.S. Department of Health, Education and Welfare Publication. C.D.C., Atlanta, Georgia.

195. Herrell, W. E. 1944. The clinical use of penicillin. J. Amer. Med. Ass. *124:* 622-627.

196. Herrell, W. E., A. Balows, and J. S. Dailey. 1955. Erythromycin in the treatment of actinomycosis. Antibiot. Med. *1:* 507-512.

197. Hertz, J. 1960. Actinomycosis. Borderline cases. J. Int. Coll. Surg. *34:* 148-168.

198. Hillman, J. D., J. van Houte, and R. J. Gibbons. 1970. Sorption of bacteria to human enamel powder. Arch. Oral Biol. *15:* 899-903.

199. Hoeprich, P. D., D. Brandt, and R. H. Parker. 1968. Nocardial brain abscess cured with cycloserine and sulfonamides. Amer. J. Med. Sci. *255:* 208-216.

200. Hogshead, H. P., and G. H. Stein. 1970. Mycetoma due to *Nocardia brasiliensis.* J. Bone Joint Dis. *52A:* 1229-1234.

201. Holdeman, L. V., and W. E. C. Moore. 1972. Anaerobe Laboratory Manual. Southern Printing Co., Blacksburg, Virginia.

202. Holm, P. 1930. Comparative studies on pathogenic anaerobic *Actinomyces.* Acta Pathol. Microbiol. Scand. Suppl. *3:* 151-156.

203. Holm, P. 1948. Some investigation into the penicillin sensitivity of human pathogenic actinomycetes. Acta Pathol. Microbiol. Scand. *25:* 376-404.

204. Holm, P. 1950. Studies on the aetiology of human actinomycosis. I. The "other microbes" of actinomycosis and their importance. Acta Pathol. Microbiol. Scand. *27:* 736-751.

205. Holm, P. 1951. Studies on the aetiology of human actinomycosis. II. Do the "other microbes" of actinomycosis possess virulence? Acta Pathol. Microbiol. Scand. *28:* 391-406.

206. Holmberg, K., and U. Forsum. 1973. Identification of *Actinomyces, Arachnia, Bacterionema, Rothia,* and *Propionibacterium* species by defined immunofluorescence. Appl. Microbiol. *25:* 834-843.

207. Holmberg, K., and H. O. Hallender. 1973. Numerical taxonomy and laboratory identification of *Bacterionema matruchotii, Rothia dentocariosa, Actinomyces naeslundii, Actinomyces viscosus,* and some related bacteria. J. Gen. Microbiol. *76:* 43-63.

208. Hotchi, M., and J. Schwarz. 1972. Characterization of actinomycotic granules by architecture and staining methods. Arch. Pathol. *93:* 392-400.

209. Howell, A. 1963. A filamentous microorganism isolated from periodontal plaque in hamsters. I. Isolation, morphology and general cultural characteristics. Sabouraudia. *3:* 81-92.

210. Howell, A., and R. J. Fitzgerald. 1953. The production of acid phosphatases by certain species of *Actinomyces.* J. Bacteriol. *66:* 437-442.

211. Howell, A., Jr., and H. V. Jordan. 1963. A filamentous organism isolated from periodontal plaque in hamsters. II. Physiological and biochemical characteristics. Sabouraudia *3:* 92-105.

212. Howell, A., H. V. Jordan, L. K. Georg, and L. Pine. 1965. *Odontomyces viscosus,* gen. nov., spec. nov., A filamentous microorganism isolated from periodontal plaque in hamsters. Sabouraudia. *4:* 65-67.

213. Howell, A., W. C. Murphy, F. Paul, and R. M. Stephan. 1959. Oral strains of *Actinomyces.* J. Bacteriol. *78:* 82-95.

214. Howell, A., and L. Pine. 1956. Studies on the growth of species of *Actinomyces.* I. Cultivation in a synthetic medium with starch. J. Bacteriol. *71:* 47-53.

215. Howell, A., and L. Pine. 1961. The classification of organisms termed *Leptotrichia* (Leptothrix) *buccalis.* IV. Physiological and biochemical characteristics of *Bacterionema matruchotii.* Bacteriol. Rev. *25:* 162-171.

216. Howell, A., and M. Rogosa. 1958. Isolation of *Leptotrichia buccalis.* J. Bacteriol. *76:* 330-331.

217. Howell, A., R. M. Stephan, and F. Paul. 1962. Prevalence of *Actinomyces israelii, A. naeslundii, Bacterionema matruchotii* and *Candida albicans* in selected areas of the oral cavity and saliva. J. Dent. Res. *41:* 1050-1059.

218. Hugh, R., and E. Liefson. 1953. The taxonomic significance of fermentative versus oxidative metabolism of carbohydrates by various gram-negative bacteria. J. Bacteriol. *66:* 24-26.

219. Hunter, G. C., and C. M. Westrick. 1957. Cervicofacial abscesses by *Actinomyces.* Oral Surg. Oral Med. Oral Pathol. *10:* 793-800.

220. Israel, J. 1878. Neue beobachtungen auf dem gebiete der mykosen des menschen. Arch. Pathol. Anat. Physiol. Klin. Med. *74:* 15-53.

221. Israel, J. 1879. Neue beitrage zu den mykotischen erkrankungen des menschen. Arch. Pathol. Anat. Physiol. Klin. Med. *78:* 421-437.

222. Israel, J. 1887. Ein beitrag zur pathogenese der lungenaktinomykose. Arch. Klin, Chir. *34:* 160-164.

223. Iwami, Y., M. Higuchi, T. Yawodi, and S. Araya. 1972. Degradation of lactate by *Bacterionema matruchotii* under aerobic and anaerobic conditions. J. Dent. Res. *51:* 1683.

224. Iyer, P. K. R., and P. P. Rao. 1971. Suspected pulmonary nocardiosis in a duck. Sabouraudia. *9:* 79-80.

225. Johnson, J. L., and C. S. Cummins. 1972. Cell wall composition and deoxyribonucleic acid similarities among the anaerobic coryneforms, classical propionibacteria and strains of *Arachnia propionica.* J. Bacteriol. *109:* 1047-1066.

226. Jones, D., and P. H. A. Sneath. 1970. Genetic transfer and bacterial taxonomy. Bacteriol. Rev. *34:* 40-81.

227. Jordan, H. V. 1971. Rodent model systems in periodontal disease research. J. Dent. Res. *50:* 236-242.

228. Jordan, H. V., and B. F. Hammond. 1972. Filamentous bacteria isolated from root surface caries. Arch. Oral Biol. *17:* 1-12.

229. Jordan, H. V., and A. Howell. 1965. Nutritional control of cellular morphology in an aerobic actinomycete from the hamster. J. Gen. Microbiol. *38:* 125-130.

230. Jordan, H. V., and P. H. Keyes. 1964. Aerobic, gram-positive, filamentous bacteria as etiological agents of experimental periodontal disease in hamsters. Arch. Oral Biol. *9:* 401-414.

231. Jordan, H. V., and P. H. Keyes. 1965. Studies in the bacteriology of hamster periodontal disease. Amer. J. Pathol. *46:* 843-857.

232. Jordan, H. V., P. H. Keyes, and S. Bellack. 1972. Periodontal lesions in hamsters and gnotobiotic rats infected with *Actinomyces* of human origin. J. Periodontol. Res. *7:* 21-28.

233. Jordan, H. V., and D. L. Sumney. 1973. Root surface caries; review of the literature and significance of the problem. J. Periodontol. *44:* 158-163.

234. Jungerman, P. F., and R. M. Schwartzman. 1972. Veterinary Medical Mycology. Lea and Febiger, Philadelphia.

235. Kanetsuna, F., and A. Bantoli. 1972. A simple chemical method to differentiate *Mycobacterium* from *Nocardia*. J. Gen. Microbiol. *70:* 209-212.

236. Kapsimalis, P., and G. E. Garrington. 1968. Actinomycosis of periapical tissues. Oral Surg. Oral Med. Oral Pathol. *26:* 374-380.

237. Kay, E. B. 1947. Bronchopulmonary actinomycosis. Ann. Intern. Med. *26:* 581-593.

238. Keir, H. A., and J. W. Porteus. 1962. The amino acid requirements of a single strain of *Actinomyces israelii* growing in a chemically defined medium. J. Gen. Microbiol. *28:* 193-201.

239. Kellenberger, E., A. Ryter, and H. Sechaud. 1958. Electron microscopic studies of DNA containing plasms. J. Biophys. Biochem. Cytol. *4:* 671-678.

240. Keyes, P. H., and H. V. Jordan. 1964. Periodontal lesions in the syrian hamster. III. Findings related to an infectious and transmissible component. Arch. Oral Biol. *9:* 377-400.

241. Keyes, P. H., and R. M. McCabe. 1973. The potential of various compounds to suppress microorganisms in plaques produced *in vitro* by a streptococcus or an actinomycete. J. Amer. Dent. Ass. *86:* 396-403.

242. Kinch, D. A. 1968. A rapidly fatal infection caused by *Nocardia caviae* in a dog. J. Pathol. Bacteriol. *95:* 540-546.

243. King, S., and E. Meyer. 1957. Metabolic and serologic differentiation of *Actinomyces bovis* and anaerobic diphtheroids. J. Bacteriol. *74:* 234-238.

244. King, S., and E. Meyer. 1963. Gel diffusion technique in antigen-antibody reactions of *Actinomyces* species and "anaerobic diphtheroids." J. Bacteriol. *85:* 186-190.

245. Kingsbury, E. W., and J. M. Slack. 1967. A polypeptide skin-test antigen from *Nocardia asteroides*. I. Production, chemical and biologic characterization. Amer. Rev. Resp. Dis. *95:* 827-832.

246. Kingsbury, E. W., and J. M. Slack. 1969. A polypeptide skin-test antigen from *Nocardia asteroides*. Sabouraudia. *7:* 85-89.

247. Kingsbury, E. W., and H. Voelz. 1968. Structural organization of the ribonucleoprotein in *Escherichia coli*. J. Bacteriol. *95:* 1478-1480.

248. Kirsch, W. M., and J. C. Stears. 1970. Actinomycotic osteomyelitis of the skull and epidural space. J. Neurosurg. *33:* 347-351.

249. Koehne, G. 1972. Isolation of *Nocardia asteroides* from a pig. Mycopathol. Mycol. Appl. *46:* 317-318.

250. Kolouch, F., and L. F. Peltier. 1946. Actinomycosis. Surgery. *20:* 401-430.

251. Kroeger, A. V., and L. R. Sibal. 1961. Biochemical and serological reactions of an oral filamentous organism. J. Bacteriol. *81:* 581-585.

252. Kumar, R., and L. N. Mohapatra. 1968. Studies on aerobic actinomycetes isolated from soil. I. Isolation and identification of strains. Sabouraudia. *6:* 140-146.

253. Kumar, R., and L. N. Mohapatra. 1968. Studies on aerobic actinomycetes isolated from soil. II. Morphological and biochemical features of the isolates. Sabouraudia. *6:* 192-202.

254. Kurup, P. V., H. S. Randhawa, and S. K. Mishra. 1970. Use of paraffin bait technique in the isolation of *Nocardia asteroides* from sputum. Mycopathol. Mycol. Appl. *40:* 363-367.

255. Kurup, P. V., H. S. Randhawa, and R. S. Sandu. 1968. A survey of *Nocardia asteroides, N. caviae* and *N. brasiliensis* occurring in soil in India. Sabouraudia. *6:* 260-266.

256. Kurup, P. V., H. S. Randhawa, R. S. Sandhu, and S. Abraham, 1970. Pathogenicity of *Nocardia caviae, N. asteroides* and *N. brasiliensis*. Mycopathol. Mycol. Appl. *40:* 113-130.

257. Kurup, P. V., and R. S. Sandhu. 1965. Isolation of *Nocardia caviae* from soil and its pathogenicity for laboratory animals. J. Bacteriol. *90:* 822-823.

258. Kurup, P. V., and J. A. Schmitt. 1973. Numerical taxonomy of *Nocardia*. Canad. J. Microbiol. *19:* 1035-1048.

259. Kwapinski, J. B. G. 1960. Researches on the antigenic structure of *Actinomycetales*. IV. Chemical and antigenic structure of *Actinomyces israelii*. Pathol. Microbiol. *23:* 158-172.

260. Kwapinski, J. B. G., E. H. Kwapinski, J. Dowler, and G. Horsman. 1973. The phyloantigenic position of nocardiae revealed by examination of cytoplasmic antigens. Canad. J. Micro-

biol. *19:* 955-964.

261. Kwapinski, J. B. G., and H. P. R. Seeliger. 1964. Immunological characteristics of the *Actiomycetales.* A review. Zentralbl. Bakteriol. Parasitenk. Infektionskr. Hyg I Ref. *195:* 805-854.

262. Lamb, J. H., E. S. Lain, and P. E. Jones. 1947. Actinomycosis of the face and neck. J. Amer. Med. Ass. *134:* 351-359.

263. Lambert, F. W., J. M. Brown, and L. K. Georg. 1967. Identification of *Actinomyces israelii* and *Actinomyces naeslundii* by fluorescent-antibody and agar-gel diffusion techniques. J. Bacteriol. *94:* 1287-1295.

264. Lambert, F. W., J. M. Brown, and L. K. Georg. 1967. Identification of *Actinomyces israelii* and *Actinomyces naeslundii* by fluorescent-antibody and agar-gel diffusion techniques. J. Bacteriol. *94:* 1287-1295.

265. Landfried, S. 1966. Immunodiffusion analysis of the soluble antigens of *Actinomyces* and anaerobic *Corynebacterium* species. M.S. Thesis, West Virginia University.

266. Landfried, S. 1972. Isolation and characterization of an antigen from *Actinomyces israelii* ATCC 12102. Ph.D. Dissertation, West Virginia University.

267. Lane, S. L., and A. H. Kutscher. 1953. Oxytetracycline in the treatment of cervical facial actinomycosis. J. Amer. Med. Ass. *151:* 986-987.

268. La Plante, E. S., E. W. Chick, and D. S. Bauman. 1967. Treatment of cervicofacial actinomycosis with ampicillin. Cutis. *3:* 739-743.

269. Leavell, U. W., D. N. Tweeddale, R. P. O'Neil, and L. R. Bryant. 1968. Pathogenesis of cutaneous lesions of actinomycosis. South. Med. J. *61:* 849-851.

270. Lechevalier, M. P. 1968. Identification of aerobic actinomycetes of clinical importance. J. Lab. Clin. Med. *71:* 934-944.

271. Lechevalier, M. P., A. C. Horan, and H. Lechevalier. 1971. Lipid composition in the classification of nocardiae and mycobacteria. J. Bacteriol. *105:* 313-318.

272. Lechevalier, M. P., and H. Lechevalier. 1970. Chemical composition as a criterion in the classification of aerobic actinomycetes. Inter. J. Syst. Bacteriol. *20:* 435-443.

273. Lechevalier, H. A., M. P. Lechevalier, and B. Becker. 1966. Comparison of the chemical composition of cell walls of nocardiae with that of other aerobic bacteria. Inter. J. Syst. Bacteriol. *16:* 151-160.

274. Lechevalier, M. P., H. Lechevalier, and A. C. Horan. 1973. Chemical characteristics and classification of nocardiae. Canad. J. Microbiol. *19:* 965-972.

275. Lee, B. Y., F. Tolete, and R. Douglas. 1970. Pulmonary nocardiosis successfully treated with chemotherapy and resection. Chest. *58:* 388-391.

276. Leers, W. D., J. Dussault, J. E. Mullens, and R. Volpe. 1969. Suppurative thyroditis: An unusual case caused by *Actinomyces naeslundii.* Can. Med. Ass. J. *101:* 715-718.

277. Lentze, F. A. 1938. Zur bakteriologie und vakzinetherapie der aktinomykose. Zentralbl. Bakteriol. Parasitenk. Infektionskr. Hyg. Abt. I. Orig. *141:* 21-36.

278. Lentze, F. A. 1961. Die Aktinomykose des Menschen. Lehrbuch der Medizinische Mikrobiologie und Infektions Krankheiten. Fisher, Stuttgart.

279. Lentze, F. 1969. Die Aktinomykose und die Nocardiosen. Die Infektionskrankheiten des Menschen und ihre Erregen. A. Gumbach ed. George Thieme, Stuttgart.

280. Lerner, P. I. 1974. Susceptibility of pathogenic actinomycetes to antimicrobial compounds. Antimicrob. Ag. Chemother. *5:* 302-309.

281. Lerner, P. I., and G. L. Baum. 1973. Antimicrobial susceptibility of *Nocardia* species. Antimicrob. Ag. Chemother. *4:* 85-93.

282. Lesher, R. J., M. A. Gerencser, and V. F. Gerencser. 1974. Morphological, biochemical and serological characterization of *Rothia dentocariosa.* Inter. J. Syst. Bacteriol. *24:* 154-159.

283. Lessel, E. F. 1960. The nomenclatural status of the generic names of the *Actinomycetales.* Int. Bull. Bacteriol. Nomencla. Taxon. Suppl. *10:* 87-192.

284. Li, Y. F., and L. K. Georg. 1968. Differentiation of *Actinomyces propionicus* from *Actinomyces israelii* and *Actinomyces naeslundii* by gas chromatography. Can. J. Microbiol. *14:* 749-753.

285. Lindenberg, A. 1909. Un nouveau mycetome. Arch. Parasitol. *13:* 265-282.
286. Listgarten, M. A., H. Mayo, and M. Amsterdam. 1973. Ultrastructure of the attachment device between coccal and filamentous microorganisms in "corn cob" formations of dental plaque. Arch. Oral Biol. *18:* 651-656.
287. Littman, L. M., and J. S. Paul. 1952. Treatment of pulmonary actinomycosis with chloramphenicol. J. Amer. Med. Ass. *148:* 608-612.
288. Loe, H. 1971. Human research model for the production and prevention of gingivitis. J. Dent. Res. *50:* 256-264.
289. Loesche, W. J., R. N. Hockett, and S. A. Syed. 1972. The predominant cultivable flora of tooth surface plaque removed from institutionalized subjects. Arch. Oral Biol. *17:* 1311-1325.
290. Lord, F. T. 1910. A contribution to the etiology of actinomycosis. The experimental production of actinomycosis in guinea pigs inoculated with the contents of carious teeth. Boston Med. Surg. J. *163:* 82-85.
291. Lord, F. T. 1910. The etiology of actinomycosis. J. Amer. Med. Ass. *55:* 1261-1263.
292. Ludwig, T. G., and H. R. Sullivan. 1952. Studies of the flora of the mouth. VIII. An examination of selected human strains of anaerobic *Actinomyces.* Aust. J. Exp. Biol. Med. Sci. *30:* 81-93.
293. Lyons, C., C. R. Owen, and W. B. Ayers. 1943. Sulfonamide therapy in actinomycotic infections. Surgery. *14:* 99-104.
294. MacCarthy, J. 1955. Actinomycosis of the female pelvic organs with involvement of the endometrium. J. Pathol. Bacteriol. *69:* 175-179.
295. MacKinnon, J. E., and R. C. Artagaveytia-Allende. 1956. The main species of pathogenic aerobic actinomycetes causing mycetomas. Trans. Roy. Soc. Trop. Med. Hyg. *50:* 31-40.
296. MacLennan, A. P. 1961. Composition of the cell wall of *Actinomyces bovis:* The isolation of 6-deoxy-L-talose. Biochim. Biophys. Acta. *48:* 600-601.
297. Macotela-Ruiz, E., and A. Gonzalez-Angulo. 1966. Electron microscope studies of granules of *Nocardia brasiliensis* in man. Sabouraudia. *5:* 92-98.
298. Macotela-Ruiz, E., and F. Mariat. 1963. Sur la production de mycetomes experimentaux par *Nocardia brasiliensis* et *Nocardia asteroides.* Bull. Soc. Pathol. Exot. *56:* 46-54.
299. Magnusson, H. 1928. The commonest forms of actinomycosis in domestic animals and their etiology. Acta Pathol. Microbiol. Scand. *5:* 170-245.
300. Magnusson, M. 1962. Specificity of sensitins. III. Further studies in guinea pigs with sensitin of various species of *Mycobacterium* and *Nocardia.* Amer. Rev. Resp. Dis. *86:* 395-404.
301. Mahgoub, E. S. 1972. Treatment of actinomycetoma with sulphamethoxazole plus trimethoprim. Amer. J. Trop. Med. Hyg. *21:* 332-335.
302. Main, J. H. P., and I. T. MacPhee. 1964. Actinomycosis of the maxilla in relation to a periodontal abscess. Oral Surg. Oral Med. Oral Pathol. *17:* 299-304.
303. Manheim, S. D., C. Voleti, A. Ludwig, and J. H. Jacobson. 1969. Hyperbaric oxygen in the treatment of actinomycosis. J. Amer. Med. Ass. *210:* 552-553.
304. Martin, W. J., D. R. Nichols, W. E. Wellman, and L. A. Weed. 1956. Disseminated actinomycosis treated with tetracycline. Arch. Intern. Med. *97:* 252-258.
305. Martinelli, B., and E. A. Tagliapietra. 1970. Actinomycosis of the arm. Bull. Hosp. Joint Dis. *31:* 31-42.
306. Mason, K. N., and B. H. Hathaway. 1969. A study of *Nocardia asteroides.* Arch. Pathol. *87:* 389-392.
307. Mathieson, D. R., R. Harrison, C. Hammond, and A. T. Henrici. 1935. Allergic reactions of actinomycetes. Amer. J. Hyg. *21:* 405-421.
308. Maurice, M. T., M. J. Vacheron, and G. Michel. 1971. Isolement d'acides nocardiques de plusiurs especies de *Nocardia.* Chem. Phys. Lipids *7:* 9-18.
309. McClung, N. M. 1960. Isolation of *Nocardia asteroides* from soils. Mycologia. *52:* 154-156.
310. McCormack, L. J., J. A. Dickson, and A. R. Reich. 1954. Actinomycosis of the humerus. J. Bone Joint Surg. *36A:* 1255-1258.
311. McGaughey, C. A., J. K. Bateman, and P. Z. MacKenzie. 1951. Actinomycosis in the dog. Brit. Vet. J. *107:* 428-430.

312. McQuarrie, D. G., and W. H. Hall. 1968. Actinomycosis of the lung and chest wall. Surgery. *64:* 905-911.

313. McVay, L. V., D. Dunavant, F. Gutherie, and D. H. Sprunt. 1950. Treatment of actinomycosis with aureomycin. J. Amer. Med. Ass. *143:* 1067-1068.

314. McVay, L. V,, and D. H. Sprunt. 1953. Treatment of actinomycosis with isoniazid. J. Amer. Med. Ass. *153:* 95-98.

315. Melville, T. H. 1965. A study of the overall similarity of certain actinomycetes mainly of oral origin. J. Gen. Microbiol. *40:* 309-315.

316. Meyen, F. J. F. 1827. *Actinomyce,* Strahlenpilz. Eine neue Pilz-Gattung. Linnaea. *2:* 433-444.

317. Meyer, E., and P. Verges. 1950. Mouse pathogenicity as a diagnostic aid in the identification of *Actinomyces bovis.* J. Lab. Clin. Med. *36:* 667-674.

318. Midwinter, G. G., and R. D. Batt. 1960. Endogenous respiration and oxidative assimilation in *Nocardia corallina.* J. Bacteriol. *79:* 9-17.

319. Miller, M. J., and P. H. Long. 1952. Successful treatment of actinomycosis with stilbamidine. J. Amer. Med. Ass. *150:* 35.

320. Mishra, S. K., and H. S. Randhawa. 1969. Application of paraffin bait technique to the isolation of *Nocardia asteroides* from clinical specimens. Appl. Microbiol. *18:* 686-687.

321. Mohapatra, L. N., and L. Pine. 1963. Studies on the pathogenicity of aerobic actinomycetes inoculated into mice intravenously. Sabouraudia. *2:* 176-184.

322. Mohr, J. A., E. R. Rhoades, and H. G. Muchmore. 1970. Actinomycosis treated with lincomycin. J. Amer. Med. Ass. *212:* 2260-2262.

323. Montgomery, R. M., and W. A. Welton. 1959. Primary actinomycosis of the upper extremity. Arch. Dermatol. *79:* 578-580.

324. Moore, W. R., and J. G. Scannel. 1968. Pulmonary actinomycosis simulating cancer of the lung. J. Thorac. Cardiov. Surg. *55:* 193-195.

325. Morris, E. O. 1954. The bacteriology of the oral cavity. V. *Corynebacterium* and gram-positive filamentous organisms. Brit. Dent. J. *97:* 29-36.

326. Mosselman, G., and E. Lienaux. 1890. L'actinomycose et son agent infecteur. Ann. Med. Vet. *39:* 409-426.

327. Munyua, W. K., and G. M. Mugera. 1970. Actinomycosis involving soft and bony tissue in swine. Bull. Epizoot. Dis. Afr. *18:* 247-251.

328. Murphy, J. B. 1885. Actinomycosis in the human subject. New York Med. J. *41:* 17-19.

329. Murray, J. F., S. M. Finegold, S. Froman, and D. W. Will. 1961. The changing spectrum of nocardiosis. Amer. Rev. Resp. Dis. *83:* 315-330.

330. Murray, I. G., and E. S. Mahgoub. 1970. Treatment of nocardiosis. Lancet. *2:* 362.

331. Myers, H. B. 1937. Thymol therapy in actinomycosis. J. Amer. Med. Ass. *108:* 1875.

332. Naeslund, C. 1925. Studies of *Actinomyces* from the oral cavity. Acta Pathol. Microbiol. Scand. *2:* 110-140.

333. Naeslund, C. 1929. Experimentelle uebertragung der aktinomykoses von menschen und versuchstiere mittels reinkulturen der actinomycesform Wolff-Israel (Act. Israeli). Acta Pathol. Microbiol. Scand. *6:* 66-77.

334. Naeslund, C. 1931. Experimentelle studien die aetiologie und pathogenesis der aktinomykose. Acta Pathol. Microbiol. Scand. Suppl. 6. *8:* 1-156.

335. Nathan, M. H., W. P. Radman, and H. L. Barton. 1962. Osseous actinomycosis of the head and neck. Amer. J. Roentgenol. Radium Ther. Nucl. Med. *87:* 1048-1053.

336. Negroni, P. 1934. Microorganismes anaerobies dans les mycétoma humains. C. R. Soc. Biol. Paris. *117:* 1239-1240.

337. Negroni, P. 1954. Micosis Profundas. Los Micetomas, Vol. 1. El Ateneo, Buenos Aires.

338. Negroni, P., and H. Bonfiglioli. 1937. Morphology and biology of *Actinomyces israelii.* J. Trop. Med. Hyg. *40:* 226-232, 240-245.

339. Neuber, E. 1940. Spezifische diagnostik und therapie der aktinomykose. Klin. Wochenschr. *19:* 736-741.

340. Nichols, D. R., and W. E. Herrell. 1948. Penicillin in the treatment of actinomycosis. J. Lab. Clin, Med. *33:* 521-525.

341. Niven, C. F., Jr., K. L. Smiley, and J. M. Sherman. 1942. The hydrolysis of arginine by streptococci. J. Bacteriol. *43:* 651-660.

342. Nocard, E. 1888. Note sur la maladie des boeufs de la Guadeloupe connue sous le nom de farcin. Ann. Inst. Pasteur, Paris. *2:* 293-302.

343. Ochoa, A. G. 1962. Mycetomas caused by *Nocardia brasiliensis* with a note on the isolation of the causative organism from soil. Lab. Invest. *11:* 1118-1123.

344. Onisi, M. 1949. Study on the *Actinomyces* isolated from the deeper layers of carious dentine. Shikazaku Zasehi. *6:* 273-282.

345. Onisi, M., and J. Nuckolls. 1958. Description of actinomycetes and other pleomorphic organisms recovered from pigmented carious lesions of the dentine of human teeth. Oral Surg. Oral Med. Oral Pathol. *11:* 913-930.

346. Orfanakis, M. G., H. G. Wilcox, and C. B. Smith. 1972. *In vitro* studies of the combined effect of ampicillin and sulfonamides on *Nocardia asteroides* and results of therapy in four patients. Antimicrob. Ag. Chemother. *1:* 215-220.

347. Ortiz-Ortiz, L., M. F. Contreras, and L. F. Bojalil. 1972. Cytoplasmic antigens from *Nocardia* eliciting a specific delayed hypersensitivity. Infect. Immunity. *5:* 879-882.

348. Overman, J. R., and L. Pine. 1963. Electron microscopy of cytoplasmic structures in facultative and anaerobic *Actinomyces*. J. Bacteriol. *86:* 656-665.

349. Paalman, R. J., M. B. Dockerty, and R. D. Mussey. 1949. Actinomycosis of the ovaries and fallopian tubes. Amer. J. Obstet. and Gynecol. *58:* 419-431.

350. Padron, S., M. Dominquez, R. Drosd, and M. J. Robinson. 1973. Lymphocutaneous *Nocardia brasiliensis* infection mimicking sporotrichosis. South. Med. J. *66:* 609-612.

351. Page, L. R., and G. N. Krywolap. 1974. Deoxyribonucleic acid base composition of *Bacterionema matruchotii*. Inter. J. Syst. Bacteriol. *24:* 289-291.

352. Panijayanond, P., C. A. Olsson, M. L. Spivack, G. W. Schmitt, B. A. Idelson, B. J. Sachs, and D. C. Nabseth. 1972. Intraocular nocardiosis in a renal transplant patient. Arch Surg. *104:* 845-847.

353. Peabody, J. W., and J. H. Seabury. 1957. Actinomycosis and nocardiosis. J. Chronic Dis. *5:* 374-403.

354. Peabody, J. W., and J. H. Seabury. 1960. Actinomycosis and nocardiosis: A review of basic differences in therapy. Amer. J. Med. *28:* 99-115.

355. Pegrum, G. D. 1964. Actinomycotic lesions in the chorio-allantoic membrane of the chick embryo. J. Pathol. Bacteriol. *88:* 323-327.

356. Petitprez, A., and J. C. Derieux. 1970. Mise in evidence de polysaccharides sur quelques types de bacteries. J. Microscopie. *9:* 263-272.

357. Pheils, M. T., D. J. Reid, and C. F. Ross. 1964. Abdominal actinomycosis. Brit. J. Surg. *51:* 345-350.

358. Phillips, M., J. R. Thurston, and A. C. Pier. 1970. Extracellular antigen of *Nocardia asteroides*. IV. Antigen fractionation with polyacrylamide electrophoresis. Amer. Rev. Resp. Dis. *101:* 545-550.

359. Pier, A. C., and R. E. Fichtner. 1971. Serologic typing of *Nocardia asteroides* by immunodiffusion. Amer. Rev. Resp. Dis. *103:* 698-707.

360. Pier, A. C., and R. F. Keeler. 1965. Extracellular antigens of *Nocardia asteroides*. I. Production and immunologic characterization. Amer. Rev. Resp. Dis. *91:* 391-408.

361. Pier, A. C., M. J. Mejia, and E. H. Willers. 1961. *Nocardia asteroides* as a mammary pathogen in cattle. I. The disease in cattle and the comparative virulence by 5 isolates. Amer. J. Vet. Res. *22:* 502-517.

362. Pine, L. 1970. Classification and phylogenetic relationship of microaerophilic actinomycetes. Int. J. Syst. Bacteriol. *20:* 445-474.

363. Pine, L., and C. J. Boone. 1967. Comparative cell wall analyses of morphological forms within the genus *Actinomyces*. J. Bacteriol. *94:* 875-883.

364. Pine, L., and L. K. Georg. 1965. The classification and phylogenetic relationships of the *Actinomycetaceae*. Int. Bull. Bacteriol. Nomencl. Taxon. *19:* 267-272.

365. Pine, L., and L. K. Georg. 1969. Reclassification of *Actinomyces propionicus*. Int. Bull. Bacteriol. Nomencl. Taxon. *15:* 143-163.

366. Pine, L., and L. K. Georg. 1974. Genus *Arachnia*. Bergey's Manual of Determinative Bacteriology. 8th ed. N. E. Gibbons, ed. Williams and Wilkins Co., Baltimore.

367. Pine, L., and H. Hardin. 1959. *Actinomyces israelii*, a cause of lacrimal canaliculitis in man. J. Bacteriol. *78:* 164-170.

368. Pine, L., H. Hardin, L. Turner, and S. Roberts. 1960. Actinomycotic lacrimal canaliculitis. Amer. J. Ophthalmol. *49:* 1278-1288.

369. Pine, L., and A. Howell. 1956. Studies on the growth of species of *Actinomyces.* I. Cultivation in a synthetic medium with starch. J. Bacteriol. *71:* 47-53.

370. Pine, L., A. Howell, and S. J. Watson. 1960. Studies of the morphological, physiological and biochemical characters of *Actinomyces bovis.* J. Gen. Microbiol. *23:* 403-424.

371. Pine, L., and J. R. Overman. 1966. Differentiation of capsules and hyphae in clubs of bovine sulphur granules. Sabouraudia. *5:* 141-143.

372. Pine, L., W. A. Shearin, and C. A. Gonzales. 1961. Mycotic flora of the lacrimal duct. Amer. J. Ophthalmol. *52:* 619-625.

373. Pirtle, E. C., P. A. Rebers, and W. W. Weigel. 1965. Nitrogen-containing and carbohydrate-containing antigen from *Actinomyces bovis.* J. Bacteriol. *89:* 880-888.

374. Poubard, J. A., I. Husain, and R. F. Norris. 1973. Biology of the bifidobacteria. Bacteriol. Rev. *37:* 136-165.

375. Poulton, E. P. 1937. Discussion on the treatment of bacterial diseases with substances related to sulphanilamide. Proc. Roy. Soc. of Med. *31:* 164.

376. Prather, J. R., C. E. Eastridge, F. A. Hughes, and J. J. McCaughan. 1970. Actinomycosis of the thorax. Ann. Thorac. Surg. *9:* 307-312.

377. Presant, C. A., P. H. Wiernick, and A. A. Serpick. 1970. Disseminated extrapulmonary nocardiosis presenting as a renal abscess. Arch. Pathol. *89:* 560-564.

378. Pritzker, H. G., and J. S. MacKay. 1963. Pulmonary actinomycosis simulating bronchiogenic carcinoma. Can. Med. Ass. J. *88:* 785-791.

379. Prolo, D. J., and J. W. Hanbery. 1968. Secondary actinomycotic brain abscess. Arch. Surg. *96:* 58-64.

380. Putman, H. C., M. C. Dockerty, and J. M. Waugh. 1950. Abdominal actinomycosis: An analysis of 122 cases. Surgery. *28:* 781-800.

381. Raich, R. A., F. Casey, and W. H. Hall. 1961. Pulmonary and cutaneous nocardiosis. Amer. Rev. Resp. Dis. *83:* 505-509.

382. Reed, M. J. 1972. Chemical and antigenic properties of the cell wall of *Actinomyces viscosus.* J. Dent. Res. *51:* 1193-1202.

383. Reed, M. J., and R. T. Evans. 1973. Cell wall peptidoglycan of *A. viscosus.* Abst. ASM *73:* 182.

384. Reed, M. J., and C. M. Gordon. 1973. Identification of soluble antigens of *Actinomyces* sp. J. Dent. Res. *52:* 88.

385. Richardson, R. L., and J. Schmidt. 1959. An oral filamentous microorganism: Cultural characteristics and microbial relationships affecting growth. J. Dent. Res. *38:* 1016-1027.

386. Ritz, H. L. 1963. Localization of *Nocardia* in dental plaque by immunofluorescence. Proc. Soc. Exp. Biol. Med. *113:* 925-929.

387. Rivolta, S. 1878. Sul cosi detto mal del rospo del Trutta e sull' *Actinomyces bovis* di Harz. Clin. Vet. Milano. *1:* 201-208.

388. Rogosa, M. 1974. Genus *Bifidobacterium.* Bergey's Manual of Determinative Bacteriology. 8th ed. N. E. Gibbons, ed. Williams and Wilkins Co., Baltimore.

389. Rose, H. D., and M. W. Rytel. 1972. Actinomycosis treated with clindamycin. J. Amer. Med. Ass. *221:* 1052.

390. Rosebury, T. 1944. The parasitic actinomycetes and other filamentous microorganisms of the mouth. A review of their characteristics and relationships, of the bacteriology of actinomycosis and of salivary calculus in man. Bacteriol. Rev. *8:* 189-223.

391. Rosebury, T., L. J. Epps, and A. R. Clark. 1944. A study of the isolation cultivation and pathogenicity of *Actinomyces israelii* recovered from the human mouth and from actinomycosis in man. J. Infect. Dis. *74:* 131-149.

392. Rosen, S. 1969. Comparison of sucrose or glucose in the causation of dental caries in

gnotobiotic rats.  Arch. Oral Biol. *14:* 445-450.

393. Roth, G. D.  1957.  Proteolytic organisms of the carious lesion.  Oral Surg. Oral Med. and Oral Pathol. *10:* 1105-1117.

394. Roth, G. D., and A. N. Thurn.  1962.  Continued study of oral *Nocardia*.  J. Dent. Res. *41:* 1279-1292.

395. Roveda, S. I. L.  1973.  Cervicofacial actinomycosis.  Report of two cases involving major salivary glands.  Aust. Dent. J. *18:* 7-9.

396. Rud, J.  1967.  Cervicofacial actinomycosis.  J. Oral Surg. *25:* 229-235.

397. Ruhrah, J.  1899.  Actinomycosis in man, with special reference to cases which have been observed in America.  Ann. Surg. *30:* 417-451, 605-631, 722-746.

398. Salmo, N. A. M., V. Rudolf, and N. T. Makki.  1969.  Actinomycosis of the colon.  Dis. Colon Rectum. *12:* 30-32.

399. Sanford, A. H.  1923.  Distribution of actinomycosis in the United States.  J. Amer. Med. Ass. *81:* 655-658.

400. Sanford, G. E., and R. O. Barnes.  1949.  Massive penicillin therapy of abdominal actinomycosis.  Surgery. *25:* 711-723.

401. Sanford, A. H., and M. Voelker.  1925.  Actinomycosis in the United States.  Arch. Surg. *11:* 809-841.

402. Savard, E. V., I. S. Snyder, and J. E. Hawkins.  1966.  Hemagglutinating antibodies to *Nocardia asteroides*.  J. Infect. Dis. *116:* 439-446.

403. Sazama, L.  1965.  Actinomycosis of the parotid gland.  Oral Med. Oral Surg. Oral Pathol. *19:* 197-204.

404. Schardt, W. M.  1956.  Corneal ulcer due to *Nocardia asteroides*.  Amer. J. Ophthalmol. *42:* 303-305.

405. Scharfen, J.  1971.  Trutnov 139/66—An unusual actinomycete combining the contradictory properties of the genera *Nocardia* and *Actinomyces*—the causative agent of submandibular mycetoma.  J. Hyg. Epidemiol. Microbiol. Immunol. *15:* 43-51.

406. Scharfen, J.  1973.  On the systematic position of microaerophilic, aerial mycelium forming actinomycetes of Trutnov 139/66 type.  Zentralbl. Bakteriol. Parasitenk. Infektionskr. Hyg. Abt. Orig. A. *225:* 95-103.

407. Scharfen, J.  1973.  Urease as a useful criterion in the classification of microaerophilic actinomycetes.  Zentralbl. Bakteriol. Parasitenk. Infektionskr. Hyg. Abt. I Orig. A. *225:* 89-94.

408. Scherp, H. W.  1971.  Dental caries: Prospects for prevention.  Science *173:* 1199-1205.

409. Schleifer, K. H., and O. Kandler.  1972.  Peptidoglycan types of bacterial cell walls and their taxonomic implications.  Bacteriol. Rev. *36:* 407-477.

410. Schmidt, J. M., and R. L. Richardson.  1962.  Serology of *Bacterionema matruchotii*.  J. Bacteriol. *83:* 584-589.

411. Schwarz, J., and G. L. Baum.  1970.  Actinomycosis.  Sem. Roentgenol. *5:* 58-63.

412. Selignan, A., J. Hantree, H. Wasserkrug, H. Dmochowski, and L. Katzoff.  1965.  Histochemical demonstration of some oxidized macromolecules with thiocarbohydrate (TCAA) or thiosemicarbazide (TSCE) and osmium tetroxide.  J. Histochem. Cytochem. *13:* 629-639.

413. Shklair, I. L., and S. Rosen.  1969.  Cariogenic potential of *Streptococcus* No. 167 in germ-free rats.  J. Dent. Res. *48:* 1313.

414. Shuster, M., M. M. Klein, H. C. Pribor, and W. Kozub.  1967.  Brain abscess due to *Nocardia*.  Arch. Intern. Med. *120:* 610-614.

415. Sibal, L. R., A. V. Kroeger, D. Kumarich, and E. Meyer.  1962.  Serological specificity of acid-soluble antigens of *Bacterionema matruchotii*.  J. Bacteriol. *83:* 811-818.

416 Slack, J. M. 1942.  The source of infection in actinomycosis.  J. Bacteriol. *43:* 193-209.

417. Slack, J. M.  1968.  Subgroup on taxonomy of microaerophilic actinomycetes.  Inter. J. Syst. Bacteriol. *18:* 253-262.

418. Slack, J. M.  1974.  Epidemiology of actinomycosis, a bacterial disease. *In* Epidemiology of Human Mycotic Diseases.  Charles C. Thomas, Springfield, Illinois.

419. Slack, J. M.  1974.  Family *Actinomycetaceae* and genus *Actinomyces*.  Bergey's Manual of Determinative Bacteriology.  8th ed.  N. E. Gibbons, ed.  Williams and Wilkins Co., Baltimore.

420. Slack, J. M., and M. A. Gerencser.  1966.  Revision of serological grouping of *Actinomyces*. J. Bacteriol. *91:* 2107.

421. Slack, J. M., and M. A. Gerencser.  1970.  Two new serological groups of *Actinomyces*. J. Bacteriol. *103:* 266-267.

422. Slack, J. M., and M. A. Gerencser.  1970.  Range of cellular and colonial morphologies of *Actinomyces israelii*.  Int. J. Syst. Bacteriol. *20:* 259-268.

423. Slack, J. M., and M. A. Gerencser.  1970.  The genus *Actinomyces*.  The *Actinomycetales*. Jena Inter. Symp. Taxon. *1:* 19-37.

424. Slack, J. M., S. Landfried, and M. A. Gerencser.  1969.  Morphological, biochemical, and serological studies on 64 strains of *Actinomyces israelii*.  J. Bacteriol. *97:* 873-884.

425. Slack, J. M., S. Landfried, and M. A. Gerencser.  1971.  Identification of *Actinomyces* and related bacteria in dental calculus by the fluorescent antibody technique. J. Dent. Res. *50:* 78-82.

426. Slack, J. M., E. H. Ludwig, H. H. Bird, and C. M. Canby.  1951.  Studies with microaerophilic actinomycetes.  I. The agglutination reaction.  J. Bacteriol. *61:* 721-735.

427. Slack, J. M., A. Winger, and D. W. Moore.  1961.  Serological grouping of *Actinomyces* by means of fluorescent antibodies.  J. Bacteriol. *82:* 54-65.

428. Slade, P. R., B. V. Slesser, and J. Southgate.  1973.  Thoracic actinomycosis.  Thorax. *28:* 73-85.

429. Smith, G.  1934.  Roentgen therapy of actinomycosis.  Amer. J. Roentgenol. Radium Ther. Nucl. Med. *31:* 823-829.

430. Smith, I. M., and A. H. S. Hayward.  1971.  *Nocardia caviae* and *Nocardia asteroides:* Comparative bacteriological and mouse pathogenicity studies.  J. Comp. Pathol. *81:* 79-87.

431. Smith, L. D., and L. V. Holdeman.  1968.  The Pathogenic Anaerobic Bacteria.  Charles C. Thomas, Springfield, Illinois.

432. Smith, W. N., and J. L. Streckfuss.  1972.  Respiratory linked dehydrogenases of *Bacterionema matruchotii*.  Abst. IADR *72:* 149.

433. Snieszko, S. F., G. L. Bullock, C. E. Dunbar, and L. L. Pettijohn.  1964.  *Nocardia* infection in hatchery-reared fingerling rainbow trout (*Salmo gairdneri*).  J. Bacteriol. *88:* 1809-1810.

434. Snijders, E. P.  1924.  Cavia-scheef-kopperij, een nocardiose.  Geneesk. Tijdschr. Ned.-Ind. *64:* 85-87.

435. Snyder, M. J., M. S. Slawson, W. Bullock, and R. B. Parker.  1967.  Studies on oral filamentous bacteria.  II. Serological relationships within the genera *Actinomyces, Nocardia, Bacterionema* and *Leptotrichia*.  J. Inf. Dis. *117:* 341-345.

436. Snyder, M. L., W. Bullock, and R. B. Parker.  1967.  Studies on the oral filamentous bacteria.  I. Cell wall composition of *Actinomyces, Nocardia, Bacterionema* and *Leptotrichia*. J. Infec. Dis. *117:* 332-340.

437. Socransky, S. S.  1970.  Relationship of bacteria to the etiology of periodontal disease. J. Dent. Res. *49:* 203-222.

438. Socransky, S. S., C. Hubersak, and D. Propas.  1970.  Induction of periodontal destruction in gnotobiotic rats by a human oral strain of *Actinomyces naeslundii*.  Arch. Oral Biol. *15:* 993-995.

439. Socransky, S. S., and S. D. Manganiello.  1971.  The oral microbiota of man from birth to senility.  J. Periodontal. *42:* 485-496.

440. Speir, N. A., J. N. Mitchener, and R. F. Galloway.  1971.  Primary pulmonary botryomycosis.  Chest. *60:* 92-93.

441. Sperling, R. L., R. Heredia, W. J. Gillesby, and B. Chomet.  1967.  Rupture of the spleen secondary to actinomycosis.  Arch. Surg. *94:* 344-348.

442. Spilsbury, B. W., and F. R. C. Johnstone.  1962.  The clinical course of actinomycotic infections.  Can. J. Surg. *5:* 33-47.

443. Stanton, M. B.  1966.  Actinomycosis of the maxillary sinus.  J. Laryngol. Otol. *80:* 168-174.

444. Stephan, R. M.  1953.  The dental plaque in relation to the etiology of caries.  Int. Dent. J. *4:* 180-195.

445. Sukapure, R. S., M. P. Lechevalier, H. Reber, M. L. Higgins, H. A. Lechevalier, and H. Prauser. 1970. Motile nocardial *Actinomycetales*. Appl. Microbiol. *19:* 527-533.

446. Susens, G. P., A. Al-Shamma, J. C. Rowe, C. C. Herbert, M. L. Bassis, and G. C. Coggs. 1967. Purulent constrictive pericarditis caused by *Nocardia asteroides*. Ann. Intern. Med. *67:* 1021-1032.

447. Sutter, V. L., and S. M. Finegold. 1972. Anaerobic Bacteriology Manual. UCLA, Los Angeles.

448. Takazoe, I. 1969. A synthetic medium for *Bacterionema matruchotii*. Bull. Tokyo Dent. Coll. *10:* 97-105.

449. Takazoe, I., and J. Ennever. 1969. Ultrastructure of *Bacterionema matruchotii*. Bull. Tokyo Dent. Coll. *10:* 45-60.

450. Tanzil, H. O. K., A. Chatim, R. Rioutomo, and H. Harun. 1972. Sensitivity of various species of *Nocardia* to rifampin *in vitro*. Amer. Rev. Resp. Dis. *105:* 455-456.

451. Taptykova, S. D., and L. V. Kalakoutskii. 1973. Low temperature cytochrome spectra of anaerobic actinomycetes. Int. J. Syst. Bacteriol. *23:* 468-471.

452. Thammayya, A., N. Basu, D. Sur-Roy-Chowdbury, A. K. Banerjee, and M. Sanyal. 1972. Actinomycetoma pedis caused by *Nocardia caviae* in India. Sabouraudia. *10:* 19-23.

453. Thiery, J. P. 1967. Mise in evidence des polysacchrides sur coupes fines on microscopie electronique. J. Microscopie. *6:* 987-1018.

454. Thompson, L. 1933. Actinobacillosis of cattle in the United States. J. Infect. Dis. *52:* 223-229.

455. Thompson, L. 1950. Isolation and comparison of *Actinomyces* from human and bovine infections. Proc. Staff Meet. Mayo Clin. *25:* 81-86.

456. Thompson, L., and S. A. Lovestedt. 1951. An *Actinomyces*-like organism obtained from the human mouth. Proc. Staff Meet. Mayo Clin. *26:* 169-175.

457. Thornley, M. J. 1960. The differentiation of *Pseudomonas* from other gram-negative bacteria on the basis of arginine metabolism. J. Appl. Bacteriol. *23:* 37-52.

458. Thurston, J. R., M. Phillips, and A. C. Pier. 1968. Extracellular antigens of *Nocardia asteroides*. III. Immunologic relationships demonstrated by erythrocyte-sensitizing antigens. Am. Rev. Resp. Dis. *97:* 240-247.

459. Tomm, K. E., J. W. Raleigh, and G. A. Guinn. 1972. Thoracic actinomycosis. Amer. J. Surg. *124:* 46-48.

460. Tsukamura, M. 1969. Numerical taxonomy of the genus *Nocardia*. J. Gen. Microbiol. *56:* 265-287.

461. Uesaka, I., K. Oiwa, K. Yasuhira, Y. Kobara, and N. M. McClung. 1971. Studies on the pathogenicity of *Nocardia* isolates for mice. Jap. J. Exp. Med. *41:* 443-457.

462. Uhler, I. V., and L. A. Dolan. 1972. Actinomycosis of the tongue. Oral Surg. Oral Med. Oral Pathol. *34:* 199-200.

463. Urdaneta, L. F., R. P. Belin, J. Cueto, and R. C. Doberneck. 1967. Intramural gastric actinomycosis. Surgery. *62:* 431-435.

464. Viroslav, J., and T. W. Williams. 1971. Nocardial infection of the pulmonary and central nervous system. Successful treatment with medical therapy. South. Med. J. *64:* 1382-1385.

465. Von Baracz, R. 1903. Report of sixty cases of actinomycosis. Ann. Surg. *37:* 336-340.

466. Walker, O. 1938. Sulphanilamide in the treatment of actinomycosis. Lancet. *1:* 1219-1220.

467. Waksman, S. A. 1950. The Actinomycetes. Chronica Botanica Co., Waltham, Massachusetts.

468. Waksman, S. A. 1959. The Actinomycetes. Nature, Occurrence and Activities. Vol. 1. Williams and Wilkins Co., Baltimore.

469. Waksman, S. A., and A. T. Henrici. 1943. The nomenclature and classification of the actinomycetes. J. Bacteriol. *46:* 337-341.

470. Wangensteen, O. H. 1932. Actinomycosis of the thorax with a report of a case successfully operated upon. J. Thorac. Surg. *1:* 612-636.

471. Wangensteen, O. H. 1936. The role of surgery in the treatment of actinomycosis. Ann. Surg. *104:* 752-770.

472. Waring, H. J. 1905. Actinomycosis of cecum, vermiform appendix and right iliac fossa. St. Bartholomew's Hosp. Rep. *41:* 197-210.

473. Weed, L. A., and A. H. Baggenstoss. 1949. Actinomycosis, a pathologic and bacteriologic study of twenty-one fatal cases. Amer. J. Clin. Pathol. *19:* 201-216.

474. Weintraub, M. I., and G. H. Glaser. 1970. Nocardial brain abscesses and pure motor hemiplegia. N.Y. State J. Med. *70:* 2717-2721.

475. Weiss, M. H., and J. A. Jane. 1969. *Nocardia asteroides* brain abscess successfully treated by enucleation. J. Neurosurg. *30:* 83-86.

476. Widra, A. 1963. Histochemical observations on *Actinomyces bovis* granules. Sabouraudia. *2:* 264-267.

477. Wilson, E. 1961. Abdominal actinomycosis with special reference to the stomach. Brit. J. Surg. *49:* 266-270.

478. Winslow, C. E. A., J. Broadhurst, R. E. Buchanan, C. Krumwiede, L. A. Rogers, and G. H. Smith. 1920. The families and genera of bacteria. J. Bacteriol. *5:* 191-229.

479. Wolff, M., and J. Israel. 1891. Ueber reincultur des *Actinomyces* und seine uebertragbarkeit auf thiere. Arch. Pathol. Anat. Physiol. Klin. Med. *126:* 11-59.

480. Wright, J. H. 1905. The biology of the microorganism of actinomycosis. J. Med. Res. *8:* 349-404.

481. Wright, L. T., and H. J. Lowen. 1950. Aureomycin in actinomycosis. J. Amer. Med. Ass. *144:* 21-22.

482. Wynn, W. H. 1908. A case of actinomycosis (streptothrichosis) of the lung and liver successfully treated with a vaccine. Brit. Med. J. *1:* 554-557.

483. Yamada, T., A. Sakai, S. Tonouchi, and K. Kawashina. 1971. Actinomycosis of the liver. Amer. J. Surg. *121:* 341-345.

484. Young, L. S., D. Armstrong, A. Blevins, and P. Leiberman. 1971. *Nocardia asteroides* infection complicating neoplastic disease. Amer. J. Med. *50:* 356-367.

485. Young, W. B. 1960. Actinomycosis with involvement of vertebral column: Case report and review of literature. Clin. Radiol. *11:* 175-182.

486. Zamora, A., L. F. Bojalil, and F. Bastarrachea. 1963. Immunologically active polysaccharides from *Nocardia asteroides* and *Nocardia brasiliensis*. J. Bacteriol. *85:* 549-555.

487. Zoeckler, S. J. 1951. Cardiac actinomycosis. Circulation. *3:* 854-858.

# Index